Critical Geopolitics

Critical Geopolitics
The Politics of Writing Global Space

GEARÓID Ó TUATHAIL

Routledge
Taylor & Francis Group

LONDON AND NEW YORK

First published 1996
by Routledge
4 Park Square, Milton Park, Abingdon, Oxon OX14 4RN
605 Third Avenue, New York, NY 10017

Routledge is an imprint of the Taylor & Francis Group, an informa business

Copyright 1996 by the Regents of the University of Minnesota

British Library Cataloguing in Publication Data

A catalogue record for this book is available from the British Library

ISBN 13: 978-0-415-15701-8 (pbk)

For Patrick Toal (1925–1994)

Contents

Acknowledgments

This book is an intellectual product of a scattered spatial journey through various educational institutions in Ireland, England, and the United States. My greatest debt is to my parents who made it possible for me to begin that journey by going to university in the first place. At Saint Patrick's College, Maynooth, Fran Walsh and others provided the necessary encouragement for me to pursue a postgraduate education in the United States. At the University of Illinois, John O'Loughlin was a supportive adviser who provided me with the freedom to pursue my interests beyond the disciplinarity of Geography at that time. At Syracuse University, I became part of a group of graduate students who were very fortunate to have John Agnew as our principal adviser. I learned a great deal of Gramscian "good sense" from John about place, politics, and history, an education that has shaped the form of this book. David Sylvan in Political Science was an inspirational guide to international relations and critical social theory. A fellowship from the Center for International Studies at the University of Southern California provided me with a stimulating environment in which to develop my initial ideas on critical geopolitics. Conversations in the company of Tom Biersteker, Anders Stephanson, James Der Derian, and others helped a great deal. In Minneapolis, the members of the Departments of Geography and Political Science at the University of Minnesota were most hospitable in extending me the opportunity to begin teaching there. Again

I owe a particular debt of gratitude to David Sylvan. In 1991–92, I had the opportunity to teach at the University of Liverpool and was enriched by the friendship of Gerry Kearns and Andy Charlesworth as well that of the students. At Virginia Polytechnic Institute and State University, I am very fortunate to have supportive colleagues in the Department of Geography and to have Timothy W. Luke, Paul Knox, and Edward Weisband as university colleagues. Tim, in particular, has been a consistent source of inspiration and encouragement. His comments, together with the detailed constructive criticism of Derek Gregory and David Campbell, made this a better book than it might have been. Thanks also to Lisa Freeman and Gretchen Asmussen for their editorial guidance.

In addition to those mentioned, I would like to thank the following for their influence and help: J. H. Andrews, Richard Ashley, Mark Bassin, Elisabeth Binder, Vincent Carey, Stuart Corbridge, Andrew Crampton, Marcus Doel, James Duncan, Brendan Gleeson, Michael Heffernan, David Livingstone, Susan Roberts, Paul Routledge, James Ryan, Michael Shapiro, Neil Smith, Peter Taylor, Michela Verardo, Rob Walker, and Cynthia Weber. A special thanks is due to Simon Dalby, who has done so much to blaze a trail for critical geopolitics, and to my long-standing friend Fintan McKenna, who once again helped me with French with his usual twisted humor. An earlier version of part of chapter 5 was previously published as "The Critical Reading/Writing of Geopolitics: Re-Reading Wittfogel, Bowman and Lacoste," in *Progress in Human Geography* 18 (1994): 313–32. It is included here by kind permission of Edward Arnold.

Introduction: Geo-Power

> *Our Geographers do not forgett what entertainment the Irish of Tir-*
> *connell gave to a map-maker about the end of the last Rebellion; for,*
> *one Barkeley [Bartlett] being appoynted by the late Earle of Devon-*
> *shire to draw a trew and perfet mapp of the north parts of Ulster (for*
> *that the old mappes were false and defective), when he came to Tir-*
> *connell, the enhabitants tooke off his head, by cause they would not*
> *have their cuntrey discovered.*
> — SIR JOHN DAVIS, SOLICITOR-GENERAL OF IRELAND, 1609[1]

Geography is about power. Although often assumed to be innocent,
the geography of the world is not a product of nature but a product
of histories of struggle between competing authorities over the power
to organize, occupy, and administer space. Imperial systems throughout history, from classical Greece and Rome to China and the Arab
world, exercised their power through their ability to impose order and
meaning upon space. In sixteenth-century Europe, the centralizing
states of the "new monarchs" began organizing space around an intensified principle of royal absolutism. In regions both within and beyond the nominal domain of the Crown, the power of royal authority over space was extended and deepened by newly powerful court
bureaucracies and armies. The results in many instances were often
violent, as the jurisdictional ambitions of royal authority met the determined resistance of certain local and regional lords. Within the context of this struggle, the cartographic and other descriptive forms of

1

knowledge that took the name "geography" in the early modern period and that were written in the name of the sovereign could hardly be anything else but political. To the opponents of the expansionist court, "geography" was a foreign imposition, a form of knowledge conceived in imperial capitals and dedicated to the territorialization of space along lines established by royal authority. Geography was not something already possessed by the earth but *an active writing of the earth by an expanding, centralizing imperial state.* It was not a noun but a verb, a *geo-graphing,* an earth-writing by ambitious endo-colonizing and exocolonizing states who sought to seize space and organize it to fit their own cultural visions and material interests.

More than five hundred years later, this struggle between centralizing states and authoritative centers, on the one hand, and rebellious margins and dissident cultures, on the other hand, is still with us. While almost all of the land of the earth has now been territorialized by states, the processes by which this disciplining of space by modern states occurs remain highly contested. From Chechnya to Chiapas and from Rondonia to Kurdistan and East Timor, the jurisdictions of centralized nation-states strive to eliminate the contradictions of marginalized peoples and nations. Idealized maps from the center clash with the lived geographies of the margin, with the controlling cartographic visions of the former frequently inducing cultural conflict, war, and displacement. Indeed, the rise in the absolute numbers of displaced peoples in the past twenty-five years is testimony to the persistence of struggles over space and place. In 1993 the United Nations High Commissioner for Refugees estimated that roughly 1 in every 130 people on earth has been forced into flight because of war and state persecution. In 1970 there were 2.5 million refugees in the world; today that figure is well over 18.2 million. In addition an estimated 24 million people are internally displaced within their own states because of conflict.[2] More recently, genocide in Rwanda left over 500,000 murdered and produced an unprecedented exodus of refugees from that state into surrounding states. Refugees continue to be generated by "ethnic cleansing" campaigns in the Balkans; economic collapse in Cuba; ethnic wars in the Caucasus; state repression in Guatemala, Turkey, Indonesia, Iraq, and Sudan; and xenophobic terror in many other states. Struggles over the ownership, administration, and mastery of space are an inescapable part of the dynamic of contemporary global politics.

It is along borders that one can best appreciate the acuteness of this perpetual struggle over space in global politics. In my own case, growing up along the border between Northern Ireland and the Republic of Ireland was an education in spatial politics, for, in this borderland, geography and politics were intensely intertwined. Whether one said "Derry" or "Londonderry," for example, in naming the second largest city in Northern Ireland was of great significance, since the prefix "London" revealed whether one acknowledged the legitimacy of the British writing of Irish space, a historically significant writing/righting that orientated Irish space around the capital of the British empire.[3] The landscape of this borderland has long been marked by signs of spatial struggles: bombed-out customs posts, cratered "unapproved" roads (periodically refilled by local residents, only to be blasted once again by the British army), and increasingly high-tech panoptic military checkpoints manned by soldiers who move only in helicopters. While peace negotiations may bring change, this frontier remains for the present a borderland ensculptured with heavily fortified checkpoints, perimeterized by invisible electronic surveillance grids, and, in certain hot spots like South Armagh, policed from above by hovering army helicopters with microwave radiation devices designed to generate electromagnetic maps of the population below. In this part of the world, geography is most conspicuously a problematic concerning the occupation and control of territorial space.

MODERNITY AND THE WRITING OF SPACE: THE GEOGRAPHIC INVENTION OF IRELAND

The problematic of the subjugation and management of space conceptualized as a territorial container requiring effective occupation by a central state apparatus first emerges in Europe in the sixteenth century. The medieval conceptualization and organization of space was religious, with maps representing the divine order of the world (Jerusalem at the center and the Mediterranean as the middle of the earth) and places organized into vertical, hierarchic ensembles (sacred places and profane places, celestial places and terrestrial places, and the like).[4] This gradually gave way to an early modern horizontal organization of space associated with ideas of state sovereignty and the emerging state system.[5] Galileo, Euclid, and Newton's representation of space as infinite, homogeneous, and absolute facilitated the rise of a modern territorial understanding of space as a horizontal order of coexistent places that could be sharply delimited and compartimentalized from each other.[6] This conception of space was eventually recognized and codified in the Treaty of Westphalia in 1648.

At the threshold of this modernity was the Elizabethan state's expansion beyond the Pale into the interiors of the island of Ireland from the middle of the sixteenth century. This expansionism provoked new forms of knowledge that sought to address the problematic of the conquest, delimitation, and mastery of space. A detailed cartography was essential in subjugating what were held to be the "wild and untamed" territories of the island. The greatest difficulty facing the armies raised by the English Crown was mastering the difficult and disorientating terrain. The woods, bogs, lakes, and mountains of Ireland concealed and sustained resistance to the jurisdictional ambitions of the English Crown. Large-scale maps were necessary for the English commanders to map out their overall strategy of conquest and subjugation, while small-scale maps were necessary for fortification schemes and to persuade an often reluctant queen to send more soldiers.[7]

Maps were also a crucial part of the technical infrastructure necessary for the plantation and governance of the space seized by the English Crown. Without maps, Ireland was an illegible surface to English planners and administrators, a disorientating space that was not yet a territory. The function of cartography was to transform seized space into a legible, ordered imperial territory. The Elizabethan plantation of Munster and the subsequent plantation of Ulster in the early seventeenth century were imperial enterprises that required considerable cartographic resources. Together with private adventurers and colonial entrepreneurs, colonial administers like Sir John Davis and Lord Mountjoy elaborated plans to make Ireland a "razed table" upon which the Elizabethan state could transcribe a neat territorial pattern.[8] In doing so they invented "Ireland" as a geographical and discursive entity. As David Barker writes:

> Not only did the Queen's administrators establish towns, lay out roads, and so on, but in a certain sense, they *made* Ireland.... As in England, what these officials needed was a demarcated terrain that they could be said to control, a specific geo-political entity that would be answerable to the discourse they brought to bear on it. Ireland

came into being as a "nation" as those who administered it marched across it, mapped it, wrote about it, and, generally speaking, produced and assembled a physical domain which, for them at least, was coextensive with the space of their own discourse. The country which resulted from their bureaucratic labours was, as much as anything else, a figure of speech — their speech.[9]

This production of "Ireland" as a geographical location went well beyond mapmaking.[10] The late Elizabethan period saw the production of a series of descriptive geographical surveys of the political condition and economic potential of the island.[11] Motivations for these surveys varied. Some sought to provide accounts of political problems and champion not only a political strategy (such as Davis's argument for a thorough colonization of the country) but their own employment prospects. Other works were associated with plantation enterprises and sought to represent Ireland as a lucrative place for commercial investments. In all, Ireland was an incomplete place in need of "civility." Observers inveighed against the strangeness of Irish customs, the paganness of Irish religious practices, and the need for a variety of improvement schemes to make the land fertile. Descriptions of the physical landscape were inseparable from governmental, commercial, and moral discourse. Dense woods were not simply dense woods, nor towns mere towns. Rather dense woods were an index of the wild and untamed nature of the Irish; they were sites of danger and potential resistance that called forth plans for the remolding of the landscape to fit English notions of order and progress. Towns were part of this cosmology of order, indicative of the advance of Anglicization and civility.[12] Irish lands were the object of a Christian (i.e. Anglican) civilizing mission. Davis remarked that "it stands neither with Christian policy nor conscience to suffer so good and fruitful a country to lie waste like a wilderness."[13] Ireland was a wild feminine land awaiting cultivation, a virgin territory in need of husbandry.[14]

Unlike the Native American encounter with the Spanish in the New World, or the British encounter with the Aborigines in Australia, the world of the Gaelic lords was not radically outside that of Elizabethan England.[15] The leader of the rebellion against the English Crown in the late sixteenth century was Hugh O'Neill, an Irish-born but English-educated lord who had previously aided the Crown's forces against other rebellious lords and who held the royal title of Earl of Tyrone. Yet, despite the relative proximity of the world of Irish lords to that of the English Crown, by the time of Davis, the preclassical episteme's rendering of "the Irish" as socially inferior to "the English" (an episteme, according to Foucault, organized on the basis of resemblance, affinity, and similarity) was giving way to a classical episteme that inscribed the "Irish" as irreducibly and permanently inferior (the classical episteme being organized on the principle of discrimination and difference).[16] From the late sixteenth century, "Ireland" was the site of an English Renaissance self-fashioning, a negative mirror of an emerging English self-image, a locational projection of negations of Englishness. The "Irish" were no longer similar to the "English" but an unassimilatable otherness that needed to be comprehensively conquered before the Irish could be reformed. "Irish nature" was innately rebellious, while "Ireland" (frequently punned as Ire-land in Elizabethan treatises) was a barbarous country that had culturally polluted those English (the Old English) who had settled there and intermarried with the native Irish. To Davis "a barbarous country must first be broken by a war before it will be capable of good government; and when it is fully subdued and conquered if it be not well planted and governed after the conquest it will eftsoons return to the former barbarism."[17] In positioning the "Irish" within the classical episteme's understanding of barbarism, they were axiomatically judged as a people without history and geography. They were thus an irritant obstacle to the progress of the one true "English" history and geography, stubborn savages blocking the production of an ordered landscape of Anglican civilization and royal harmony.

GEOGRAPHY AND GOVERNMENTALITY

Bartlett's decapitation is at the threshold of a modernity that was to distinguish itself by its ambitious redrafting of space around the principles of empire and state sovereignty (which were, in the case of Elizabethan England, the same thing). Crucial to the imposition and, as time went on, the increasingly intensified functioning of this modernity were those governmental apparatuses that produced the territorialization of space and also the (Euclidian-Galilean) spatialization of territory. These governmental practices and the art of government insinuated within them sought, with varying and uneven degrees of force and reason, to impose ordered visions of space, territory, and geography upon ambivalent lands, terrains, and cultures. The genealogical conundrum they mark is a problematic I wish to term *geo-power*, the functioning of geographical knowledge not as an innocent body of knowledge and learning but as an ensemble of technologies of power concerned with the governmental production and management of territorial space. The problematic of geo-power concerns the modern governmentalization of geography from the sixteenth century onward, a time, as Foucault notes, when government as a general problem demanding public and intellectual thought explodes in terms of its relevance and significance.[18]

Foucault's elaboration of the question of "governmentality" in his 1978–79 course, "Security, Territory and Population," is suggestive in helping specify this problematic, but it should be remembered that Foucault's arguments are somewhat speculative and exploratory. In his "Governmentality" lecture, Foucault seeks to specify the broad contours of a genealogy of theories of the art of government. The question of government is not simply a question of the government of a state by a prince but a question of the government of personal conduct and the government of souls and lives. Foucault seeks to differentiate a narrow medieval and princely understanding of the art of government, which took the sole interest of the prince as its object of rationality (most pointedly in Machiavelli's *The Prince*), from a more ambitious but nevertheless somewhat immobilized introduction of economy (household management) into the art of government. Here the notion of government is extended both in an upward (the person who wishes to govern the state must first learn how to govern himself or herself) and downward (a well-run state should promote well-run families and properly behaved individuals) direction. Government is "the right disposition of things, arranged so as to lead to a convenient end" (93).[19] Foucault traces the origins of this art of government to the sixteenth century and "the whole development of the administrative apparatus of the territorial monarchies." It was also "connected to a set of analyses and forms of knowledge which began to develop in the late sixteenth century and grew in importance during the seventeenth, and which were essentially to do with knowledge of the state, in all its different elements, dimensions and factors of power, questions which were termed 'statistics,' meaning the science of the state" (96). Finally, this art of government, which brought into existence what Foucault later calls an "administrative state," also encompasses the practices of mercantilism and the Cameralist's science of police.

The development of a rational art of government from the sixteenth century, however, was somewhat immobilized and stifled by the persistence of monarchical conceptions of sovereignty. It is not until the early eighteenth century, according to Foucault, with the emergence of the problem of population, that the art of government recenters itself around the theme of economy not as family management but as the management of the population of the state (that is, economics in a modern sense). Prior to the emergence of population, Foucault claims, "it was impossible to conceive the art of government except on the model of the family, in terms of economy conceived as the management of a family; from the moment when, on the contrary, population appears absolutely irreducible to the family, the latter becomes of secondary importance compared to population, as an element internal to population" (99). What came into being in the eighteenth century, Foucault suggests, was a "governmental state" that defined the population as a datum, as a field of intervention, and as an objective of governmental techniques. "Political economy" was the science and objective of the governmental(ized) state.

Foucault's genealogy of the art of government and the forms of the state is exceedingly sketchy and sweeping in its conception. Nevertheless, bringing together theories of the art of government with the development of governmental apparatuses and forms of knowledge designed to develop knowledge of the state provides us with a framework to specify geo-power as a historical problematic of state formation. In specifying this problematic, Foucault's own observations on the changing place of space and territory in the history of the art of government are underdeveloped and potentially confusing. For example, Foucault's contrast between the "government of territory" (which he associates with the juridical sovereignty that concerned Machiavelli) and the "government of men and things" is somewhat misleading. The more specific contrast he strives to make is between a narrowly juridical conception of territory (a state of divinely ordained jurisdiction) and a broadly materialistic and resourcist conception of territory (a state effectively occupied and mapped for wealth, colonization, and security). As he himself notes, the administrative state is, among other things, concerned with "territory with its specific qualities, climate, irrigation, fertility, etc." (93). This changing vision of the territory of the state required permanent bureaucratic technologies of power that could map, describe, catalog, inventory, order, and arrange the "things" of government. Often, Foucault suggests elsewhere, the city was the model for the reordering and rearranging of the territory of the state.[20]

Also somewhat misleading is Foucault's suggestion that the governmental state "is essentially no longer defined in terms of its territoriality, of its surface area, but in terms of the mass of its population with its volume and density, and indeed also with the territory over which it is distributed, although this figures here only as one among its component elements" (104). But again the point is not simply that territoriality is no longer important but that its conceptualization and instrumentalization within the practices of government have changed. The historical dimensions and forms of that change—the differing arts of government practiced historically by the likes of Thomas Cromwell (circa 1485–1540), Anne-Robert-Jacques Turgot (1727–81), or Louis Napoleon Bonaparte (1769–1821)—are only now getting the sustained study they deserve. Each of these figures represents an interesting moment in the history of geo-power. Thomas Cromwell, Henry VIII's principal secretary, is situated at the threshold of the earliest form of the administrative state, the expansionist and centralizing state form that characterized the "new monarchies" of the sixteenth century.[21] Turgot, a French physiocrat and reputedly the person who first used the term "political geography," was an innovative state administrator at a time when the administrative state of the ancien régime was collapsing because of the contradictions of France's absolutist monarchy.[22] The failure of his fiscal reforms during his brief tenure between August 1774 to May 1776 as controller-general for Louis XVI marked a crucial stage in the demise of the ancien régime. Turgot's work anticipates the demise of the logic of an administrative state beholden to absolutism and the beginnings of a governmental state. Napoleon's regime marks the fully fledged expression of the governmental state, a state where geography played a crucial role as a scientific technology aiding not only military conquest but also the gathering of statistics for rational government by The Great State.[23]

STUDYING GEO-POWER

Arguably, the problematic of geo-power in England began with the reforms of Thomas Cromwell during the English Reformation. Certainly the territorial ambitions of the Elizabethan state and Bartlett's decapitation was a consequence of an initial phase in the historical development of geo-power. He was the first of many geographers who would be murdered because their profession was a technology of power in the service of a centralizing and imperialistic state.[24] His death helps crystallize some methodological points that are worth establishing at the outset concerning the relationship between geography and power.

First, geography is not a natural given but a power-knowledge relationship. For Foucault, power and knowledge directly imply one another: "There is no power relation without the correlative constitution of a field of knowledge, nor any knowledge that does not presuppose and constitute at the same time power relations."[25] As a general methodological principle, we should not approach the histories of forms of knowledge as if knowledge can only exist outside the injunctions, demands, and interests of power relations. Rather, histories of knowledge should be situated within the historical and geographical context of the power relations that not only were there when the form of knowledge emerged and consolidated itself as a system of statements and procedures but that also helped give the form of knowledge its very birth and existence. "Far from preventing knowledge, power produces it."[26] The very identity of the subject position "geographer," "cartographer," or "geopolitician," for example, the specification of geographical/cartographic/geopolitical objects to be recognized and represented by this subject, and the "geographical techniques" by which these objects are arranged, presented, and projected are all effects of power-knowledge relations.

Second, the general power relations within which to situate modern geographical knowledge is the centralization and imperialist expansion of the modern European state system across the globe from the sixteenth century onward. Centralization and imperialist expansionism required new aggressive forms of geographical power/

knowledge to supervise the seizure and disciplining of space. Imperialism, as Edward Said has noted, is after all "an act of geographical violence through which virtually every space in the world is explored, charted, and finally brought under control."[27] The battle between the Gaelic power structure in Tirconnell and the English Crown was a conflict over sovereignty generated by an expansionist state encroaching upon the hitherto relatively unchallenged local autonomy of regional lords.[28] Bartlett was appointed by an Earl of the Elizabethan state, and his mission was to redraft the space of Tirconnell so the region could be incorporated within the realm of this state. His job was to geo-graph the earth of Ulster, to make its terra incognito an imperial territory so its resistant foreignness could be eradicated and replaced by a recognizably Elizabethan space.

Given the contemporary politics of Ireland, I should note here that the conflict between the last of the Gaelic lords in the late sixteenth century and the Elizabethan colonial apparatus was never a straight conflict between a clear and unambiguous Irish and English identity. The Irish nationalist argument of a discrete Irish island being invaded by a foreign English Other is a determinist form of geographical reasoning that falsely equates national identity with island geography and ignores the fluidity of Ireland's borders and social system. As the historian Roy Foster notes, the word "foreign" was applied to the "English" enemy by various "Irish" lords at this time but it could also be used for anyone who threatened one's lands. "Localist, aristocratic archaism was a more dominant motif of the Gaelic consciousness in the early seventeenth century than anything like a national sense of identity."[29] The inside and outside of Ireland were never firmly established and beyond dispute. The very formation, consolidation, and maintenance of geographically specified identities is precisely what was in contestation in the north of Ulster at the time of Bartlett's murder. It was not a question of a struggle between fully formed states but of the historical and geographical struggles to impose state sovereignty. The politics of geo-graphing, then, does not begin as a problematic of geography and *international politics* but a problematic of the attempt to produce international politics geographically. Rather than assuming a fully formed state system and state-delimited identities, my concern is the power struggle between different societies over the right to speak sovereignly about geography, space, and territory. To begin with the geographical identities created and fostered by the state system is to fall into what John Agnew has called the "territorial trap."[30] It is to forget the historical struggles that went into the creation and maintenance of states as coherent territories and identities seeking international legitimacy.[31] It is to operate with a frozen, ahistorical model of the state that fails to appreciate forms of spatiality (organizations of space) other than that of an idealized territorial state presumed to have a discrete inside and outside.

The third methodological point concerns institutionalized ways of seeing and displaying space. Bartlett's activity on the colonial frontier in Ulster was only the beginning of the process of geographical violence that was to spread across the rest of the globe in the following centuries. As containers of a fledgling modernity, the expansionist new monarchies of the sixteenth century were slowly, unevenly, and erratically (depending on the state in question) imposing a general perspectivalist vision of space and a neutral conception of time upon the territories they incorporated and annexed. As Timothy Luke has argued, the state sought to establish its power by *in-state-ing* itself in space. It sought to secure its authority as both the omniscient illustrator and omniscient narrator of territory.[32] Space was homogenized (Euclidianized) and measured from a central point, which was normally the seat of government or royal authority. This central point constituted the fixed spectatorial position from which panoramic visions of official state territory were constructed. Ulster was provincialized (made into a province) in this manner.[33] In historical practice, of course, these panoramic visions of territory differed considerably as the techniques and technologies for displaying space in a perspectivalist manner evolved. Bartlett, whom Davis, revealingly, felt obliged to specify as an officially appointed authority, produced maps of the province of Ulster that look, to today's eyes, like shaded drawings (see Figure 1).[34] Nevertheless, these were shaded visions of space that sought to establish an English military gaze over the provincialized space of Ireland. They sought to render this troublesome region visible, to site it as imperial space.

But Bartlett lost his eyes and head. His murder, because the inhabitants would not have their country discovered, highlights a fourth aspect of geo-power: resistance. In seeking to keep their territory a blank space on the maps of the Elizabethan state, the chiefs of Tirconnell were actively resisting the drive to systematize and codify a

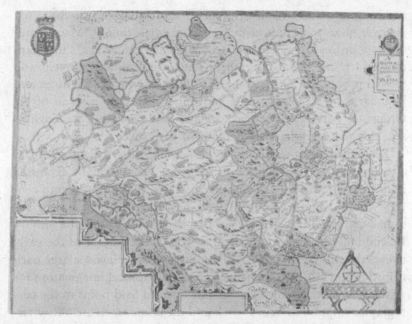

Figure 1. Geo-Power. Richard Bartlett's 1602 map of Ulster. By kind assistance of the Board of Trinity College, Dublin.

singular imperial geography at the expense of multiple local geographies. In traditional Gaelic bardic culture, according to Foster, "the terrain was studied, discussed and referenced; every place had its legend and its own identity. *Dindsenchas,* the celebration of place-names, was a feature of this poetic topography; what endured was the mythic landscape, providing escape and inspiration."[35] By the early seventeenth century, however, this Gaelic bardic culture was largely destroyed. With the effective subjugation of the indigenous population and power centers of the island came an intensified cartographic textualization and expropriation. The Cromwellian campaign in Ireland was followed by the Down Survey of Sir William Petty. First commissioned in 1654, its purpose was to facilitate the distribution of forfeited lands to Cromwellian soldiers and those who had given money for the army's support. Still, large parts of Ireland resisted mapping by the administrative apparatus of the English Crown. It was not until the Ordinance Survey of 1824 that detailed and comprehensive six-inch maps of the island were undertaken, being finally completed just before the famine in 1846. In Anglicizing and co-opting Gaelic names, the survey was smoothing out the struggle over geography that began in Elizabethan times. "Masquerading as a process of systematic record," one critic notes, "the mapping of Ireland was a prolonged act of cultural displacement and textual processing, in the course of which ancient place-names and boundaries were incorporated and reinscribed." As a result of the Ordinance Survey, "an official Ireland was produced, an English-speaking one, with its own ideology of Irish space."[36]

Resistance to the writ of the imperial state, nevertheless, remained a central feature of Irish history, a resistance that frequently expressed itself in anti-imperialist geographical terms. By the nineteenth century we can begin to speak of a modern Irish nationalist imagination, a contested but largely Catholic nationalist vision of a unitary Ireland independent of British rule. The Gaelic revival of the early twentieth century, in which William Butler Yeats played a large part, helped invent a heroic anti-imperialist geographical imagination for this nation. Both Seamus Deane and Edward Said point to the cartographic impulse in Yeat's work and the effort to create a new territoriality outside the imperialist frame in literatures of anti-imperialist resistance.[37] Said describes this as the "search for authenticity, for a more congenial national origin than that provided by colonial history, for a new pantheon of heroes, myths, and religions."[38] Although there were many ironies in this attempt to write against a hegemonic imperial history and geography, it was nevertheless a refusal to be represented in accordance with the official imperial vision of time and space. The decolonization struggle in Ireland was never only a struggle to decolonize physical space and territory but also to decolonize identity, history, and geographical knowledge.[39]

Reflecting on the general problematic of empire, geography, and culture, Said has remarked that "none of us is completely free from the struggle over geography."[40] This struggle, as he and others suggest, is not a straightforward struggle over the naming of places or the ownership of land, important though these issues are. Nor is it solely a matter of soldiers and cannons, with mapmakers in train, busily textualizing the landscape, carving it up and subjugating it to a mathematical grid designed as the basis for a plantation or envisioned community of order, loyalty, and civility. The struggle over geography is also a conflict between competing images and imaginings, a contest of power and resistance that involves not only struggles to represent the materiality of physical geographic objects and boundaries but also the equally powerful and, in a different manner, the equally material force of discursive borders between an idealized Self and a demonized Other, between "us" and "them." Viewed from the colonial frontier, geography is not just a battle of cartographic technologies and regimes of truth; it is also a contest between different ways of envisioning the world.[41]

GEO-POWER AND GEOPOLITICS

By the late nineteenth century, a new horizon of geo-power was surfacing as the last pockets of unclaimed and unstated space were surrounded and enclosed within the colonizing projects of expansionist empires and territorializing states. As the Eurocentrically imagined blank spaces on the globe succumbed to the sovereign authority of governmental institutions and imperial science, the surface of the globe appeared for the first time as a system of "closed space," an almost completely occupied and fully charted geographical order. The dawning of this new order of space, together with the transformative effects of technological change on the exercise of imperial power across space, provoked the emergence of a distinctive genre of geo-power within the capitals of the Great Powers. The name this new genre of geo-power acquired was "geopolitics."

The term "geopolitics" is a convenient fiction, an imperfect name for a set of practices within the civil societies of the Great Powers that sought to explain the meaning of the new global conditions of space, power, and technology. It names not a singularity but a multiplicity, an ensemble of heterogeneous intellectual efforts to think through the geographical dimensions and implications of the transformative effects of changing technologies of transportation, communications, and warfare on the accumulation and exercise of power in the new world order of "closed space." Like other forms of geo-power, these writings were governmentalized forms of geographical knowledge, imperial rightings from an unquestioned center of judgment that sought to organize and discipline what was increasingly experienced as unitary global space into particularistic regimes of nationalistic, ideological, racial, and civilizational truth. Circulating within the developing media of civil society (which ranged from elite markets for scholarly books to the yellow journalism of jingoistic newspapers), these discourses were motivated attempts to frame the spectacle and flux of the new global political scene within the terms of imperialistic and militaristic agendas, agendas actively cultivated and pushed by political, economic, and bureaucratic interest groups within the state-societies of the Great Powers.

If the term "geopolitics" is a convenient fiction, it is also an inconvenient one, for it is a sign overloaded with many different meanings even in the critical discourse of academia. Among many critical scholars, geopolitics has become an appealing and handy summary term for the spatiality of modernity as a whole. Michael Shapiro, for example, describes the institutionalization of state-centric discourse as "modern geopolitical discourse," a discourse that silences "the historical process of struggle in which areas and peoples have been pacified, named, homogenized, and fixed in modern international space."[42] Rob Walker is even more general when he writes of "how theories of international relations manage to constrain all intimations of a chronopolitics within the ontological determinations of a geopolitics, within the bounded geometric spaces of here and there."[43] Among critical geographers, geopolitics has been inflated to describe the most general dimensions of modernity. John Agnew and Stuart Corbridge, for example, suggest that modern geopolitical discourse begins in the encounters between Europeans and others during the so-called Age of Discovery, when a modern/backward and European/ non-European schema for dividing up the world came into use.[44] Simon Dalby suggests something abstractly similar in claiming that "the essential moment of geopolitical discourse is the division of space into 'our' place and 'their' place; its political function being to incorporate and regulate 'us' or 'the same' by distinguishing 'us' from 'them,' 'the same' from 'the other.' "[45]

The difficulty with such generalized inflations of the concept of geopolitics, however, is that they can efface the historical and geographical particularity of geopolitics as a way of envisioning and writing space-as-global from the turn of the century. The term "geopolitics" was first coined in 1899. As a consequence of the imagined significance of a German school of geopolitics in explaining Nazi foreign policy during World War II, geopolitics became the name of a tradition with a canon of classic texts and a parade of prophetic men. After the war, the term's relative and forced coherence dissolved, although it retained great symbolic significance for certain right-wing thinkers in civil society and certain military academies from South Africa to South America. It had symbolic significance in that geopolitics had the cachet of a coherent materialist approach to international politics, an approach that named itself in suitably masculinized terms as "hardheaded." For the Right, geopolitics was a politically correct, anti-Marxist materialism that had its foundations in the permanent realities of the earth. It was figuratively rooted in an imagined earth of natural laws, eternal binary oppositions, and perpetual struggles against dangerous rivals. In certain institutions and circumstances—among military officers in Brazil, Argentina, and Chile in the 1950s, for example—geopolitical thinking was de rigueur. Out of this training in geopolitics developed national security state doctrines that underpinned the murderous activities of bureaucratic authoritarian regimes in Latin America over three decades. In Chile, Augusto Pinochet, a professor of geopolitics at the Chilean War Academy, led a successful military coup, overthrowing the state's democratically elected government, and installed himself as military dictator and president. The U.S. government that encouraged this coup was likewise enamored with geopolitics. In the Nixon White House of the early 1970s and the Reagan White House of the early 1980s, geopolitics was also de rigueur, although what was meant by geopolitics differed from administration to administration. Because of Kissinger's popularization of the term, geopolitics spiraled well beyond the so-called geopolitical tradition to become a synonym for the space of global politics. The inflation of the term in recent critical intellectual discourse, therefore, is understandable and not new.

My tactical resistance to this inflation is not motivated by a desire to recover an original or stable meaning for the term but by a desire to recall and respect its complex and often unacknowledged genealogical history. Geopolitics is not a concept that is immanently meaningful and fully present to itself but a discursive "event" that poses questions to us whenever it is evoked and rhetorically deployed. It is a problematic best approached historically and contextually, a problematic concerning the writing of the global that requires an antiglobal(izing) method of inquiry that avoids treating "it" as a stable and singular "it," a linear and smooth historical surface for theoretical work. Because of this, I have composed a book of interventionist essays rather than a systematic study of an immanent and unitary concept. In calling this book *Critical Geopolitics,*

I am not naming a new critical position to oppose an old reactionary one but rather opening up a complex and somewhat neglected problematic for analysis. While each of the chapters can be considered discrete and self-standing essays, they are arranged to intersect with and build upon one another. All share a general concern with the entwining of governmentality and geographical knowledge in the writing of global space. All explore the politics of geo-graphing global political scenes. They have common sources of intellectual inspiration (Foucault mostly, but sometimes Derrida) and recurrent themes of interest (governmentality, power/knowledge, and ocularcentrism).[46] The resultant book is not a panoramic vision or panoptic survey of a singularity called "geopolitics" but a set of con-textual explorations of the problematic imperfectly marked by the term "geopolitics."

Chapter 1 introduces the general context, key intellectuals, and discursive strategies that characterize the so-called classical geopolitical tradition. In this chapter I begin to develop an immanent critique of this tradition, which I subsequently refine in chapter 2 into a general problematization of the spatialization of global politics by intellectuals of statecraft. The phenomenon of the geopolitical tradition, I suggest, forces us to confront the general question of how global space is produced and organized by governmentalizing intellectuals of statecraft. How is "international politics" written spatially? How is it envisioned geographically? How is it made visible as a spatial spectacle? How is it seen and scened, and to what end? These are the questions raised in the first two chapters and subsequently pursued in the following chapters. Pressed with critical vigor, the phenomenon of geopolitics discloses a broader problematic that I term *geo-politics*, the politics of writing global space. Critically scrutinizing the textuality of that which is presented to us as "geopolitics" leads us to problematize the pervasive geographical politics of foreign policy discourses, the ways in which the global political scene is geo-graphed by foreign policy regimes of truth.

In disclosing a broad problematic of geo-politics from an interrogation of geopolitics, it is not my wish to use the hyphen as the mark of an absolute difference between a manifest geopolitics, on the one hand, and a latent and ubiquitous geo-politics, on the other hand. Geopolitics is permanently hyphenatable. As chapters 3, 4, and 5 demonstrate, whenever we seek "geopolitics" we end up encountering not a distinct and stable object but different historical condensations of geo-politics. Building upon a discussion of the British geographer Halford Mackinder in chapter 1, chapter 3 considers the writings of Mackinder in some detail and seeks to deconstruct the way in which he sought to institutionalize a British imperial geopolitics as the very form and raison d'être of the discipline of geography in the early twentieth century. Chapter 4 reviews the emergence of "geopolitics" as a discursive object in wartime America and seeks to explicate the imaginative economy that came to be written around it as a dangerous but desirable practice at this time. Chapter 5, the longest in the book, reviews various historical attempts to write a meaning around "geopolitics" in a critical manner. Such a review illustrates not only that the critical investigation of geopolitics has a long history but that the efforts to round up and capture geopolitics within a critical cage often lead to an essentializing of the concept and a concomitant blindness to questions of con-textuality.

Chapters 6 and 7 move beyond an engagement with the history of the concept of geopolitics to investigate the practical geo-politics of U.S. foreign policy discourse. Chapter 6 considers how a new location on the global scene of the early 1990s—Bosnia—came to be written in U.S. foreign policy between 1991 and mid-1994. Chapter 7 addresses the efforts of two U.S. intellectuals of statecraft (Edward Luttwak and Samuel Huntington) to rescript the global political scene amid the vertigo of postmodernity, a vertigo precipitated for them by the end of the Cold War, compounded by globalization and intensified by global media vectors dissolving the distance between the near and the far, subjects and objects, spectators and spectacles, thus imperiling the very conditions of possibility of the modern spatial scene. Both chapters scrutinize the politics of the governmentalizing will to organize global political space into particular visions of order and disorder, geo-political visions I attempt to question, displace, and disrupt.

There is no concluding chapter, for my goal is not to bring the problematic of geopolitics to closure in the traditional sense. This book is only a start on the road toward a richer genealogy of geopolitics as a twentieth-century constellation of geo-power. It seeks to sketch out the journey and make a few specific historical road stops, stops that I recognize are limited in that they are largely confined to the Anglo-American realm. Many other stories of American, Argentinean, Brazilian, Chilean, Chinese, French, German, Italian, Japanese, and Russian geopolitics remain to be told, as do the fascinating flow-mations of the new geopolitics of the post–Cold War world. All the essays in this book are works of geopolitics with a radical agenda. They seek to disturb the innocence of geography and politicize the writing of global space. As such, they make common cause with all those who have resisted and continue to resist the exercise of geo-power by those centers of modern authority who wish only to make the world in the image of their maps.

1

Geopolitics

Geography does not argue; it just is. — NICHOLAS SPYKMAN[1]

The region of knowledge that would later be dubbed "geopolitics" was born in the colonial capitals of the rival empires of the late nineteenth century, within the established universities, geographical societies, and centers of learning of the Great Powers. From 1870 onward the Great Powers of Europe embarked upon an unprecedented program of imperial expansionism and territorial acquisition, a clamor for overseas colonies that the United States, its own continental empire building complete, joined by the end of the century. The scramble for Africa, for example, gave Europe thirty new colonies and protectorates, 10 million square miles of new territory, and 110 million new subjects.[2] This was a period that saw the growth of imperialist institutions and associations in both political and civil society dedicated to propagandizing the increasingly enfranchised masses into believing colonial expansionism was in everyone's interest. It also saw the institutionalization of geography as a self-fashioned "scientific" discipline in universities and, more important, in national secondary-school systems of imperial Germany, France, Great Britain, the United States, and other "lesser powers" like Italy and Japan.[3] In this new context, a stratum of organic intellectuals of empire emerged within the educational institutions of the Great Powers, many of whom explicitly addressed and theorized the influence of geography

21

on the social evolution of states and the conduct of foreign policy. Intellectuals like Alfred Mahan and Nicholas Spykman in the United States, Friedrich Ratzel and Karl Haushofer in Germany, Rudolf Kjellen in Sweden, and Halford Mackinder in Great Britain helped codify a mode of reasoning about international affairs that would, in context of World War II, come to be organized and categorized as constituting a "geopolitical tradition."

While the notion of a "geopolitical tradition" should always be approached with caution, it does group together this set of male intellectuals from quite different national backgrounds and cultures into a common tradition of thought on international affairs for certain ostensibly good reasons. First, those intellectuals associated with geopolitics before World War II (sometimes called "classical geopolitics") were invariably imperialists of one sort or another. Although some, like Rudolf Kjellen, did not live in active imperialist states, all sought to promote an imperialist agenda within the political culture of their own state and adopted "national" culture. This imperialist agenda (which includes the anticolonial imperialism of certain American intellectuals) presupposed the superiority of their own national variant of European or Western civilization. It also presupposed the superiority of the white race over other races and of nationally organized capitalism over other forms of the organization of economic life, like transnational capitalism or communism. The development of a strong, united, and efficient empire was considered a necessary precondition of world power and a desirable activity for modern ambitious states to pursue. To promote this imperialist agenda more effectively, a number of these theorists entered the political process or sought out positions of influence as public intellectuals and governmental advisers within their states.

Second, one intellectual current shared by early geopolitical thought and underwriting its white supremacist assumptions was neo-Lamarckism. Commonly known as "Social Darwinism," Lamarckian biology is actually quite distinct from Darwinian biology in its emphasis on design rather than randomness in the evolutionary process. Neo-Lamarckian biology argued that organisms are directly modified by their environment, whereas Darwin held that evolutionary processes are influenced by chance variations in species. It was distinguished by, first, a doctrine of the inheritance of acquired characteristics and, second, a belief that the directive forces of organic variation were will, habit, and environment. Because it speeded up the whole evolutionary process and placed its emphasis on social and environmental factors, neo-Lamarckism proved to be an amenable "scientific" justification for the superiority of the white European races and the naturalness of imperialism (among other things).[4] Because the environment played such a key directive role in its conceptualization of evolution, the historian of geographical thought, David Livingstone, has argued that neo-Lamarckism was "crucial to the project of carving out some cognitive space in academia for geography as a scholarly discipline," a proposition he fully documents in reviewing the writings of Halford Mackinder and Friedrich Ratzel.[5]

Third, what made it possible to retroactively invent a geopolitical tradition with Mackinder and Ratzel as its "founding fathers" was the development and codification, within their writings, of a distinctive way of approaching international politics. Like many other intellectuals in the late nineteenth century, these geographers approached the scene of the world with a natural attitude, a philosophical approach to reality grounded in Cartesian perspectivalism.[6] This approach sought to establish the foundations of knowledge on the basis of Cartesian cogito rationalism (of which there were many competing variants after Descartes). In simple terms, the world is taken to be a reality that exists "out there," separate from the consciousness of the intellectual. The basis of this attitude is the Cartesian divide between an inner self and an outer reality, between an internal mind and an external world of objects. On the outside is a preexistent world, flooded with light and surrounding the self on all sides; on the inside is a domain of pure mental functioning tied to a perceptual apparatus upon which the luminous scene of the surrounding world makes an impression. The relationship of the intellect to the world, therefore, is that of a viewing subject and a viewed object. Perception is the name for the process by which the passive specular consciousness of the viewing subject experiences the stream of informational stimuli from the outside world of dazzling surfaces bathed in light.

The gaze of the observing subject upon the world is represented by the Cartesian perspectivalist tradition as a neutral and disembodied gaze. In the classic Albertian formulation of perspective (derived from the notebooks of Leonardo), the organ of perception is a monocular eye, a single eye removed from the rest of the body and suspended in diagrammatic space. The visual field is two-dimensionalized, with the suspended eye witnessing, not interpreting. There is no need to process the stimuli from the outside world, since they "possess an intelligibility fully formed and theirs by virtue of the inherent intelligibility of the outer world."[7] The challenge for science (whether it be biology, geology, or geography) is to record the already nascent transparency of the outer world in as comprehensive and essentialist a manner as possible.

This privileging of the sense of sight in systems of knowledge constructed around the ideal of Cartesian perspectivalism promoted the simultaneous and synchronic over the historical and diachronic in the explanation and elaboration of knowledge. Within science, the natural attitude sanctioned forms of knowledge concerned with the synchronic unfolding and display of essential forms and patterns. It facilitated the subordination of history to space and fostered the spatialization of perceived phenomena rather than their historization. This spatialization of objects and events in the world reduced history to a set of timeless essences, conflicts, and universal laws. The world is enframed and pictured as unchanging. History is an affair on the surface of nature, the true substratum of which the steadfast mental gaze of the observer seeks to objectify. Contrary to the arguments of Edward Soja, space was not subordinated or invisible in Western intellectual thought at the fin de siècle.[8] The frame of the emergent social sciences — where society was equivalent to the territory of the state — and of political discourse more generally (busily inventing or renegotiating the limits of the nation or empire as imaginary community) was inescapably spatial. Geopolitics, a phenomenon too easily marginalized by Soja (37), was not an aberration but part of mainstream political discourse that was always conscious of the "instrumentality of space."

THE GEOPOLITICAL GAZE AND THE THEATER OF INTERNATIONAL POLITICS

Classical geopolitics (to the extent that this names an under-specified unity) has its origins in the emerging geographical conditions of world order at the end of the nineteenth century. In the wake of the tremendous scramble among the Great Powers for their own imperial real estate on the African continent, the surface of the world political map appeared for the first time to be relatively occupied. All the major powers (and some minor ones as well) had staked their claim on territory that had appeared blank and empty on European maps only a few decades beforehand. The scramble for empty space was at an end; the struggle for relative efficiency, strategic position, and military power among the competing imperial systems entered a new phase.

What precisely the particular character of that new phase would be was very much a matter of debate. In an attempt to represent the geographical and political meaning of this new phase, a hybrid discourse combining self-scientizing geographical discourse searching for universals and speculative governmental discourse on imperial strategy in a new world order emerged in the texts of intellectuals like Halford Mackinder. Later consolidated as something distinctive under the sign "geopolitics," this hybrid discourse was not particularly extraordinary at the time but part of a general wave of fin de siècle theorizing on the new century that had begun in the 1880s. Driven by an expanding market for books among the socially aspiring middle classes, this mode of discourse sought to explain in scientific terms the direction in which history was moving. To an ambitious intellectual like Mackinder, the governmentalizing of geographical discourse so that it addressed the imperialist dilemmas faced by Britain in a post-scramble world order was a splendid means of demonstrating the relevance of his "new geography" to the ruling elites of the state. What Mackinder and others meant by the "new geography" we will examine in greater detail in chapter 3. What he helped establish and codify at the beginning of the new century was a distinctive geographical gaze upon international politics, a gaze I wish to examine in detail here. Since Mackinder's January 25, 1904, address to the Royal Geographical Society, "The Geographical Pivot of History," is generally considered to be a defining moment in the history of geopolitics, a text to which histories of geopolitics invariably point, let us start by examining it.[9]

The date of this address is significant for a number of reasons. A few years earlier Mackinder had undertaken an ascent of Mount Kenya in order to establish his credentials as a manly, exploring geographer. By the early twentieth century, however, unexplored space was hard to find and there did not seem to be much of a future for geography as exploration and discovery. In 1902 Sidney and Beatrice

Webb founded the Co-Efficients (national efficiency and collective solutions) Dining Club as a forum for discussing and promoting imperial reforms. Mackinder, together with others like Sir Edward Gray, Bertrand Russell, H. G. Wells, and Leo Amery, was active in the debates on seapower, imperial defense, social reform, and preferential trade in the club (which dissolved in 1908). The Boer War had finally ended that year, a war that proved to be the longest, costliest, bloodiest, and most humiliating war for Britain between 1815 and 1914.[10] The logistical nightmare of raising and supplying an army to fight in South Africa had made a strong impression on Mackinder, who became director of the London School of Economics in 1903. In May of that same year, in addition, Joseph Chamberlain, colonial secretary since 1895, launched his campaign for a system of "imperial preferences" within the British Empire, in effect a drive to create a tariff wall around the empire against outside powers. It was a cause that was to dominate British politics for the next three years and to which Mackinder was to rally; Chamberlain provoked him not only to reevaluate his previous commitment to free trade but also to reconsider the position of the British Empire within the transformed geographical conditions of the new century.

The background to Mackinder's address to the RGS in late January 1904 was thus one of debate concerning the transformed conditions of the British Empire and the need to reform its structure. As an active advocate of the empire's modernization, Mackinder begins by specifying what he sees as a fundamental change in international politics, a change that requires new forms of intellectual practice by geographers and other intellectuals:

> Of late it has been a common-place to speak of geographical exploration as nearly over, and it is recognized that geography must be diverted to the purpose of intensive survey and philosophic synthesis. In 400 years the outline of the map of the world has been completed with approximate accuracy, and even in the polar regions the voyages of Nassen and Scott have very narrowly reduced the last possibilities of dramatic discoveries. (421)

The world is, in other words, at the end of geography as the "science" of discovery and exploration. With the last remaining unknown regions of the globe effectively unveiled by European eyes, the definitive outlines of a Eurocentrically calibrated world map are emerging. Geography has reached the culmination of its planetary circumferential mission. It has triumphed. The nature of geographic activity and inquiry, as a consequence, must change. What is now possible is "intensive survey" and "philosophic synthesis," two activities that up until this time were only possible at the regional and not the global scale because the map of the world was incomplete. With the conquest and complete appropriation of even the most remote regions of the world, however, the struggle for unmarked and unsettled space was coming to an end. The four hundred years of European spatial expansionism were described by Mackinder as "the Columbian epoch." Now a new epoch was beginning:

> From the present time forth, in the post-Columbian age, we shall again have to deal with a closed political system, and none the less that it will be one of world-wide scope. Every explosion of social forces, instead of being dissipated in a surrounding circuit of unknown space and barbaric chaos, will be sharply re-echoed from the far side of the globe, and weak elements in the political and economic organism of the world will be shattered in consequence. There is a vast difference of effect in the fall of a shell into an earthwork and its fall amid the closed spaces and rigid structures of a great building or ship. Probably some half-consciousness of this fact is at last diverting much of the attention of statesmen in all parts of the world from territorial expansion to the struggle for relative efficiency. (422)

Note that Mackinder's identification of a Columbian and a post-Columbian epoch specifies time in terms of space. The division is a product of the spatialization of history, the reduction of the complex and heterogeneous emergence of the modern world system to spatially defined categories that have a supposed innate transparency.

The distinctiveness of Mackinder's gaze is built upon the claim that the space of the world is now, for all intents and purposes, known, occupied and closed. The world has become a single unified globe of occupied space, a system of closed space (a closed spaceship earth) where events in one part inevitably have their consequences in all other parts. It is no longer possible to treat various struggles for space in isolation from one another, for all are part of a single worldwide system of closed space. The world of international interactions is now global.

Such, at least, is what Mackinder suggests. Underpinning this argument is an unquestioned geographical ethnocentrism, an ethnocentrism that initially sustained the belief that certain regions of the world political map were blank, undiscovered, and unoccupied and is now proclaiming the world political map a closed system of space, settled, occupied, and named.[11] The barbaric chaos of unknown space has been replaced by the civilizational certainties of the scientifically unveiled, one and only true geography of the planet. In not recognizing that geography is as much a political domination and cultural imagining of space, Mackinder's "end of geography" thesis could be described as a triumphalism blind to its own precariousness, although, of course, it was precisely to alert the ruling class of the empire to the precariousness of the British hold on global space that he spoke. What Mackinder does not anticipate (or, more accurately, what he wishes to prevent by his advocacy of imperial reform) is that the British, European, and, more generally, Western domination of global political space would be challenged profoundly in the new century. The struggle over geography had only begun. In interpreting the "end of geography" as a diversion from the struggle for territorial expansion to the struggle for relative efficiency among imperial states, Mackinder was oblivious to those who came to define it as the struggle for territorial and cultural independence.

Mackinder's thesis that every explosion of social forces reechoes across the globe and that weak societies are destroyed in the process was arguably true ever since the Europeans began plundering the New World and South American specie fueled inflation in seventeenth-century Europe. The capitalist world economy at that time, of course, was not global, although it became so, according to Wallerstein, in the early twentieth century, leaving all places on the planet (including the polar regions) touched by its transformative dynamics.[12] It is what Mackinder makes of the implications of this globalization for the practice of geography within the context of a relative-efficiency rivalry among empires that leads him to specify a new global approach to the world of international politics. What the post-Columbian "end of geography" condition has wrought is not simply the need for a "new geography" (see chapter 3) but, more important, the possibility of a new and powerful way of seeing, a way of seeing that can be the basis for the construction of a "new geography." For the first time, certain worldwide patterns and processes that had previously remained hidden are becoming visible. A worldwide view is now possible, a view that is not only global in a geographical sense but in an explanatory sense also.[13] The key statement of this new worldwide qua global gaze is the following:

> It appears to me, therefore, that in the present decade we are for the first time in a position to attempt, with some degree of completeness, a correlation between the larger geographical and the larger historical generalizations. For the first time we can perceive something of the real proportion of features and events on the stage of the whole world, and may seek a formula which shall express certain aspects, at any rate, of geographical causation in universal history. If we are fortunate, that formula should have a practical value as setting into perspective some of the competing forces in current international politics. (422)

Here we have a Cartesian perspectivalist organization of space with a detached viewing subject surveying a worldwide stage whose intelligibility is, for the first time, becoming visible and transparent. The relationship between the viewer and this worldwide stage is akin to that of a removed observer watching a theatrical production or a spectator viewing a panorama. Both of these metaphors are particularly important. As the profusion of theatrical metaphors throughout the geopolitical tradition attest, geopolitics can be described as the production of "international politics" as theater. Geopolitics is production in an early modern sense of the word. As Jean Baudrillard has noted, the original sense of "production" is not to materially manufacture but to render visible and make appear (producere).[14] To produce is to set everything up in clear view so it can be read, can become real and visible. Production is associated with completeness and totality. Theater, likewise, is associated with access to the order of the whole. Theater as a glass or mirror to the greater world, Stephen Daniels and Denis Cosgrove note, was a common metaphor for revealing order in the macrocosm.[15] In geopolitical narratives, what is being produced in the more complete post-Columbian condition is an "international politics" conditioned but not determined by the previously hidden forces of physical geography in combination with technologies of mobility and relative geographical position.

The metaphor of the panorama is interesting, for panoramas, paintings in the form of a cylinder looked upon by a viewer from the center, were particularly popular in the nineteenth century before the advent of moving pictures. Lieven De Cautier has suggested that many of the inventions that characterize the organization of space in modernity (encyclopedias, museums, world exhibitions, grand system philosophy, great novel cycles, evolution) are deeply panoramic. All are inspired by the belief that we can collect and classify things by putting them together along one horizon.[16] What is most important about all these activities is that they are global in ambition. They strive toward globality qua totality.

Mackinder's geopolitical gaze is a "setting into perspective" of the competing forces of current international politics. This setting into perspective is, first, panoramic in that it offers a *tour d'horizon* of "the stage of the whole world." It seeks to render the dramatic spectacle of international affairs visible in an all-encompassing global way. It is a widening of vision, a broadening of focus beyond the regional or continental scale. This setting into perspective is, second, a steadying of vision, a deliberate construction of a perspectivalist triangle of vision with the sovereign monocular eye/I at its base and the stage/spectacle/scene of history or international politics at the far wall of the triangle. The fixed disembodied eye of the geographer qua strategist is its foundation. Finally, this setting into perspective is also a distantiation from the immediate surface of appearances of "international politics," from the to and fro of normal events, from the confusing stimuli of everyday life. This act of distantiation, however, is not an elimination of stimuli but a controlled production of the stimuli of the panorama. Commenting after Mackinder's address, the Conservative politician Leo Amery noted that "it is always enormously interesting if we can occasionally get away from the details of everyday politics and try to see things as a whole, and this is what Mr Mackinder's most stimulating lecture has done for us tonight. He has given us the whole of history and the whole of ordinary politics under one big comprehensive idea."[17]

Given this, it makes sense to speak about the mise-en-scène of the geopolitical gaze, the art of setting a stage scene or arranging a pictorial representation. The mise-en-scène concerns the mechanism of theatrical illusion, "the relationship between what the audience sees and the staging and framing that produces it."[18] In Mackinder's case, the mechanisms of illusion of his own stimulating lecture–theater show are the geographicalization of history through narrative and visual maps. In his narrative, the "scene" comprises two different stage domains or stage levels. On the one hand, we have a physical/

climatological/geographical/material/spatial/natural (back)ground to the stage while, distinct from this, we have a human/historical/temporal/political/cultural foreground or surface of appearances. The universal relationship between these two stage domains is the producer's declaration that "Man and not nature initiates, but nature in large measure controls" (422).

Out of the drama of man versus nature, Mackinder scripts the spectacle of "international politics" as a three-scene conflict (pre-Columbian, Columbian, and post-Columbian) between the permanent protagonists of east and west, the Roman and the Greek, Europe and Asia, land power and sea power, with transformations in technologies of mobility providing each scene with a distinct look.[19] In the pre-Columbian scene, the balance of power lay with land power. Asia, the seat of land power, dominated Europe and shaped its history by a series of horse-borne invasions (together with the Vikings, who are marginalized by Mackinder because they do not fit into this first land-power-in-the-ascendancy scene). In the Columbian scene, however, the balance of power lay with sea power. European navies expanded overseas and turned the tables on Asia by encaging it. In the post-Columbian scene, however, the balance threatens to swing back to Asia, where railways are transmuting the conditions of land power and threaten to permit the continental resources of the heartland to be used as the basis for an empire of the world.

Mackinder supplements this production of global space by using a series of maps, which were probably shown as lantern slides at the original address. The first of these is a framed map of eastern Europe's physical geography before the nineteenth century, the second the political divisions of eastern Europe at the time of the third crusade, and the third the political divisions of eastern Europe after the accession of Charles V.[20] From his address, it is clear that Mackinder considers the first map to be the crucial physical setting upon which the changing political divisions portrayed in the second and third maps are based. This physical setting, however, is not orographic but a product of climate and vegetation; the key contrast is between forest and marsh country and steppe country.

Mackinder's final map is "The Natural Seats of Power," which features a Mercator-projection world enframed in an oval shape to simulate (badly) sphericality (see Figure 2). This is perhaps the most

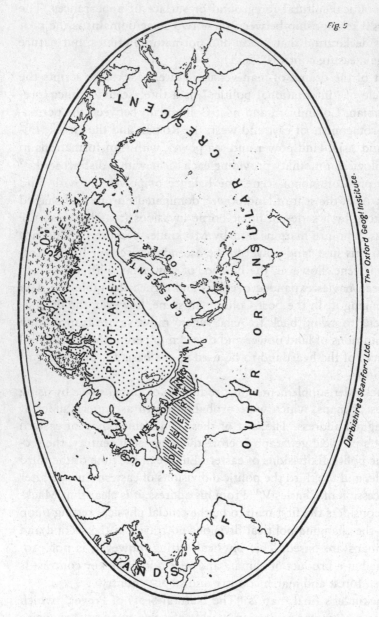

Figure 2. Unveiled by the geopolitical gaze. The Natural Seats of Power. From Halford Mackinder, "The Geographical Pivot of History," The Geographical Journal 23 (1904): 121. By kind assistance of the Royal Geographical Society.

famous map in the geopolitical tradition, a classic part of the mise-en-scène of geopolitics. The map is meant to create the illusion of the unveiled geographical determinants of power. It is the global vision the post-Columbian made possible, the pure panorama produced by the penetrative gaze of the global voyeur. Yet, interestingly, the map is not purely visual but is marked with text. Huge swaths of territory are stamped with a definitive positionality and function: "Pivot Area," "Inner or Marginal Crescent," and "Lands of the Outer or Insular Crescent." These are the macrogeographical identities around which Mackinder spatializes history and reduces it to formulaic equations and timeless east/west, land/sea conflicts. "The social movements of all times," he concluded, "have played around essentially the same physical features" (437).

Mackinder's address before his audience at the Royal Geographical Society can be described as theater about theater. His purpose was to describe the back stage that enframed the historical foreground, the pivots upon which transitory scenes and spectacles turned. His lecture was a show in which he demonstrated to his lecture-theater audience "those physical features of the world which I believe have been most coercive of human action." His aim was to "exhibit human history as part [in a phrase revealing the neo-Lamarckian panorama he had in mind] of the life of the world organism" (422). Mackinder's geography qua geopolitics (for they are the same in his work) is a theatrical production of the illusion of order in an all-encompassing three-scene spatializing spectacle called "The Geographical Pivot of History."

The development and codification of a distinctive geographical qua geopolitical gaze by Mackinder and others is undoubtedly part of a broader struggle over the reorganization of the experience of seeing in the late nineteenth and early twentieth centuries. In counterdistinction to the "primitive" spatiality of colonial natives, the unsettling, experimental spatiality of the avante-garde, or even the everyday (s)pace of the swirling modern imperial capital, the "founding fathers" of geography chose to organize the new discipline around an imperial vision of space, an ordered vision Derek Gregory, after Timothy Mitchell, has termed the production of the "world-as-exhibition."[21] A generalized feature of colonial culture, the staging of the world-as-exhibition (discussed further in chapter 2) was deeply implicated in power relations of many different types (along axes of race, gender, and class) besides the overtly political, nationalist, and colonial. An intensification of already existent tendencies in the objectification of the world by self-fashioning sites of power and authority, the production of the world-as-exhibition is, for Gregory, "the cartography of objectivism, which claimed to disclose a fundamental and enduring geometry underlying the apparent diversity and heterogeneity of the world."[22] In distinction from previous regimes of objectification, however, it was strongly committed to a visual and aesthetic organization of the space of the world and to the dream of a "total view" of global space, to panoramic display of all that there is in the world.

Implicated in the operation of this panoramic vision was the panopticon, the impulse to *master* (and the masculinist dimensions of this term should not be lost on us) all that is exhibited. The panopticon, as is well known, also arranged objects into a single encompassing horizon but did so in order to monitor and control the objects thrown into definition by the gaze. This impulse is also to be found in the geographical productions that seek to arrange and display the world around a sovereign center of judgment, a rational imperialist I/eye that observes and in so doing disciplines the ambiguity, contingency, and (barbarian) chaos of international affairs and the world. It not only represents itself as perceiving "something of the real features and events on the stage of the whole world" but also defines its mission as the reduction of these features and events to formulas of causation. The drama of international politics is not only on display but also under surveillance by imperial eyes that seek to impose a normative system of control and intelligibility upon its indeterminacy. In shedding light upon the scenes of international politics, the geopolitical gaze is also forcing them into particular prisons of meaning. The geopolitical envisioning of the global scene is inseparable from the desire to use the displayed scene for one's own purposes. International politics is a struggle for relative efficiency, from Mackinder's imperialist point of view, and the quality of one's ability to visualize the global scene is part of that struggle.

Overall, Mackinder's struggle to establish a fixed point of view was not only a struggle against other imperialist states but against an intensifying modernity and fin de siècle time-space compression that was eroding the possibility of establishing a fixed point of view. Martin Jay describes the late nineteenth century as a period of the crisis of the ancient scopic regime of Cartesian perspectivalism, as improvements in communication, artistic movements (Impressionism, Dadaism, Cubism), and new technological developments like cinema decentered and distracted the imperial gaze.[23] When he later came to write about the teaching of geography and history, Mackinder railed against the "great and insidious danger" represented by the "modern facilities for pictorial teaching and amusement," which inhibited proper focus with their vast multiplicity of distracting images. "It is to be feared," he wrote, "that the confusion will be worse confounded by the present youthful debauchery in the penny picture-palace."[24] Given his overall uncomfortableness with modernity, it is no stretch to argue that his elaboration of a panoramic and panoptic geopolitics was a conscious reaction to the tumult and turmoil of the fin de siècle period.[25] His modernism was a modernism of reaction, which sought to place Cartesian perspectivalism at the center of the new discipline of geography. Six years after Mackinder's 1904 address, the practical dominance of this Cartesian perspectivalism in everyday life was grandly proclaimed to be at an end. "On or about December, 1910," Virginia Woolf boldly proclaimed, "human character changed."[26] In his history of space, Henri Lefebvre concurred, declaring that around 1910 the space of common sense, of social practice, and of political power and everyday life was shattered.[27] But, as he astutely adds, commonsense space, Euclidean space, and perspectivalist space did not disappear in a puff of smoke but remained within our consciousness, knowledge, and educational methods. The geopolitical gaze, born in conditions of time-space compression and fin de siècle turmoil, had a future among those elites who required the spinning world to be disciplined by a fixed imperial perspective.

OTHER PRODUCTIONS OF GLOBAL SPACE

So far we have only discussed the geopolitical productionism of Halford Mackinder. To get a brief sense of other productions of the global scene and other dimensions of the geopolitical gaze, I shall quickly review the main ideas of certain key intellectuals who came to be considered "geopoliticians" later in the twentieth century, so-called seers of the geographical foundations of history. Obviously I cannot investigate their geopolitical schemas in any detail here. My purpose is simply to introduce these figures because they are part of the geopolitics story (they reappear in subsequent chapters) and to begin to problematize the gaze they deployed to organize the production and scripting of global space.

Friedrich Ratzel (1844–1904) and the Biologization of Global Space

A founding figure and towering intellectual of the discipline of geography in Germany, Friedrich Ratzel began his academic career with a doctorate in zoology in 1868. His first book was a popularization of the ideas of Darwin, which in Germany were to take the form of a neo-Lamarckism known as *Darwinismus*.[28] Forced into journalism to make a living, he became a travel correspondent for a series of moderately liberal newspapers who occasionally supported conservative causes and were generally deferential to the Bismarckian state. Writing assignments throughout the world, including a lengthy stay in the United States in the 1870s, earned Ratzel a reputation as a geography specialist, but it was the decision of the Bismarckian state to institutionalize the discourse of geography in the German university system that earned him an academic job in Munich in 1875 in the new discipline. As he later recalled, "I returned from America and they told me that they needed geographers. Thus, I collected and coordinated all the facts I had observed and collected on Chinese immigration to California, Mexico and Cuba and wrote my work on Chinese immigration, which was my practical thesis."[29] A decade later he was awarded a professorial chair at the University of Leipzig and taught there until his death in 1904.

The son of a senior domestic servant, Ratzel began his political career as a radical liberal but steadily moved to the right, becoming, as his academic career flourished, a leading theorist and propagandist for a conservative agrarianism that believed independent peasants (of which it had a romantic, idealized view) were the bedrock of the German *Volk*.[30] Ratzel's interest in questions of migration was both academic and political. His first major work, *Anthropo-Geographie* (1882), argued that the most successful peoples are those who are forever expanding into new regions and taking them over, impressing themselves and their culture (conceptualized as an array of collective traits) upon nature to create distinctive cultural landscapes. The organization of superior traits by certain peoples is the basis of human progress and cultural evolution, the key carriers of which are independent peasant agriculturalists. This diffusionist anthropology and geography dovetailed with Ratzel's political advocacy from the 1870s onward of a migrationist colonialism as the most desirable form of German imperialism.[31]

Ratzel's writing proceeded from a "scientific" method derived from natural history and zoology, a non-hypothesis-testing positivism centered on a kingly Cartesian subject (in the manner of Alexander von Humboldt) observing and comparing discrete objects in space. From a wide collection of observations and comparisons, Ratzel proceeded to formulate certain historical laws governing the diffusion of living entities, be they plants, peoples, or states. In *Political Geography* (1897), he outlined what are claimed to be the natural laws governing the territorial expansion and enlargement of states. The state, according to Ratzel, "is a living organism, and therefore cannot be contained within rigid limits—being dependent for its form and greatness on its inhabitants, in whose movements, outwardly exhibited especially in territorial growth or contraction, it participates."[32] For Ratzel, "each people located on its essential fixed area, represents a living body which has extended itself over a part of the earth and has differentiated itself either from other bodies which have similarly expanded by boundaries or by empty space."[33] Ratzel postulated seven general laws: (1) the size of the state grows with its culture; (2) the growth of states follows other manifestations of the growth of peoples, such as level of agricultural development and broadness of their geographical horizons; (3) the growth of the state proceeds by the annexation of smaller members into an aggregate; (4) the boundary is the peripheral organ of the state, the bearer of its growth as well as its fortifications, and takes part in all of the transformations of the organism of the state; (5) the state strives toward the envelopment of politically valuable positions as it grows; (6) the first stimulus to the spatial expansion of states comes from outside the state organism; and (7) the general tendency toward territorial annexation and amalgamation is transmitted from state to state and continually increases in intensity.

As an organism in a competitive struggle for existence, every great state with a growing population needed space in order to sustain and nourish its civilization. This need of states for ever increasing space was described by Ratzel as the need for lebensraum, or "living-space."[34] The struggle for living space between different cultures was the engine of progress and change in human history. States with large territorial spaces and vibrant colonizing civilizations were, according to Ratzel, the wave of the future. *Grossraum* (large space) states like the United States, Russia, and China were destined to become world powers. In order to secure world power itself, Germany must join the race for lebensraum. Since Europe was already a crowded continent and Germany's territorial base relatively small, the only way it could obtain lebensraum was by overseas territorial expansion in Africa. Ratzel actively campaigned for this policy in various colonial advocacy leagues and conservative nationalist organizations in the 1880s. After Bismarck's fall he was heavily involved in organizing the Pan-German League and later the Navy League. In the early years of the twentieth century, he campaigned with other academics (the so-called fleet professors) for the establishment of a world-class German navy as a means of furthering the quest for lebensraum. By this time, according to Smith, Ratzel had come to see the European continent, specifically a Germanic *Mitteleuropa,* as the most appropriate site for German imperialist efforts.[35] More than anything else, Friedrich Ratzel provided a political vocabulary with an aura of science for the German Right, a vocabulary that articulated and justified an exteme nationalist desire for space that was to precipitate two worldwide wars in the twentieth century.

Alfred Mahan (1840–1914) and the Strategic Gaze

Ratzel's advocacy on behalf of an extensive German fleet was undoubtedly propelled by the writings of Alfred Mahan, the U.S. evangelist of sea power in the late nineteenth and early twentieth centuries whose influence on the thinking of Kaiser Wilhelm II and Admiral Tripitz, among others, is well known. Born into a military family (his father was a professor at West Point), Captain Alfred Thayer Mahan's life as a relatively undistinguished U.S. naval officer and neophyte lecturer at the floundering Naval War Academy in Annapolis was transformed by the popular success of his second book, *The Influence of Seapower upon History, 1660–1783* (1890) and the international celebrity status he acquired because of it. A presidential adviser, Anglophile, imperialist, egoist, historian, naval reformer, social Darwinist, theologian, and teacher, Mahan was prototypical of a caste of intellectuals that would become much more numerous in the twentieth century: the geopolitical expert, the international affairs commentator for hire. Mahan's initial success with *The Influence of Seapower upon History* was the beginning of a lucrative second career for him, a career he pursued with vigor for ideological reasons but also for financial gain. In a detailed portrait of Mahan, Robert Seager II describes how financial calculations intimately shaped what Mahan decided to write once he became a literary star. Between 1879 and his death in 1914, Mahan wrote 137 articles, most of them quick and superficial "potboilers" for middle-class current affairs magazines like the *Atlantic, Collier's Weekly,* the *Forum, Harpers Monthly, North American Review,* and the *Times* (London). Mahan was also particularly enterprising in negotiating serialization deals on his more scholarly books and in converting his scattered articles into books. By the time he died, he had published twenty books, including a multivolume trilogy on the general thesis of the influence of seapower upon history.

That Mahan's name had such sign value that magazines were willing to pay him handsomely for his views was due to the worldwide reputation he gained as the author of *The Influence of Seapower.* The international success of this book was a consequence of a number of factors. First, the book was designed and marketed as a popular work of grand history with a timely thesis. Mahan later described his formulation of that thesis — that control of the sea was crucial to the course of history and the prosperity of states — as a moment of divine revelation, a perception that "turned inward darkness into light — and gave substance to shadow." This thought, which he described as coming "from within," as dawning upon his "own inner consciousness," he claimed he owed "to no other man."[36] Yet Mahan's thesis was hardly new, stretching back as it did to Themistocles, Thucydides, and Xenophon.[37] The section that generated the greatest comment and speculation among American and British readers and reviewers, Seager notes, was not the five-sixths of the book that dealt with straight history but the introductory chapter entitled "Elements of Sea Power," which was added at the last minute in an effort to make the whole work "more popular."[38] In this section, Mahan outlines six basic conditions affecting the development of seapower by states, six conditions that mirror exactly those found in an article by a young naval officer by the name of William Glenn David published in the Naval Institute's *Proceedings* in 1882, an article Mahan almost certainly had read. Mahan made these the centerpiece of "his thesis." These conditions were (1) the geographical position of a state vis-à-vis the sea; (2) the physical features of a state in relation to the seas, the length of its coastline, and the number, depth, and protected nature of its harbors; (3) the extent of its territory and the relationship of physical geography to human geography; (4) the number of its population; (5) the commercial-mindedness or otherwise of the national character; and (6) the character of the government, the operational distinction being between despotic states (Carthage, Spain) and democratic states (England, the United States).

Ronald Carpenter notes another reason for the success of Mahan's book: his rhetorical strategy.[39] Like many later twentieth-century works of geopolitics, Mahan's book had a cachet of expertise and authority (although the book is based completely on secondary sources and is factually incorrect in places). Mahan cultivated a plain, direct style that sought to illustrate a "few general principles" and "large, plain, simple ideas" designed to be "suggestive to the man in the street."[40] This belief that a few general principles could explain the whole history of sea power and indeed warfare was rooted in Mahan's religious commitment to what Seager describes as a "God-centered mechanistic universe surfeited with order, pattern, and intelligence, a universe filled with divine truths gradually being revealed by an infinite God to finite men through favored historians (such as himself) who identified and popularized the central themes that were manifestations of God's existence and handiwork."[41] Mahan's rhetorical style was appealing because he wrote history as revelation. A few general principles could explain the whole history of naval warfare. They provide what Mahan's reviewers praised as his "keen eye for discovering what is permanent"; his "rare insight," which was "penetrating"; and the "convincing truths" he brought "into array."[42] Like Mackinder's, Mahan's "insights" became pithy generalizations and epigrammatic statements that lent his writings an aura of sagacity and expertise. The very phrase "sea power," a term Mahan reputedly coined, was part of this rhetorical economy of persuasion, a concept with a transhistorical cachet yet graspable essence.

But perhaps the most socially significant reason for the success of Mahan's book and his subsequent celebrity is that he advocated a general policy that was already part of the agenda of various factions of national ruling elites at the time, namely naval militarism. In the United States, Theodore Roosevelt and other reformers were pushing for the construction of a Great Fleet as an instrument of territorial and commercial expansionism, a cause to which Mahan provided crucial intellectual impetus and legitimacy. In Great Britain in 1894 during his last naval command aboard the *Chicago,* Mahan was wined and dined by the elite of British society, including Queen Victoria and her grandson Kaiser William II. Both Oxford and Cambridge conferred honorary degrees, while the *Times* pronounced him the "new Copernicus," for he had done for naval history what Copernicus had done for astronomy.[43] Yet Mahan was being lionized not simply for his intense Anglophilia, an Anglophilia that sprung from his belief in the White Man's Burden, the "Yellow Peril," and a racial hierarchy with English-speaking Anglo-Saxons at the top; as Seager notes, the British academic, political, and naval establishments trotted Mahan out, "cheered him lustily for their own purposes (an increase in naval appropriations), and politely deposited him back on the inhospitable decks of the *Chicago.*"[44] Mahan was a useful figure because he aided the cause of domestic naval militarism. In Germany, Kaiser William II was extravagant in his praise for Mahan, also instrumentalizing his work to further his own naval militarist program. In Japan, where all of Mahan's work was translated and taught in the Naval Academy at Etajima, the story was the same.[45] Revealingly, none of these states followed Mahan in his arguments against *Dreadnought*-like "big gun" battleships.

For our purposes, the most interesting aspect of Mahan's work is the cultivated "strategic" gaze Mahan brought to bear upon the historical record of naval warfare. Both the geopolitical and strategic gaze are similar in that they both are variants of a Cartesian perspectivalism that subordinated the heterogeneous disorder of the historical to an imagined homogeneous order of the spatial. Michel de Certeau noted the importance of the spatialization or flattening out of history in the functioning of the general concept of strategy.[46] Strategy, he suggests, is associated with a Cartesian attitude; "it is an effort to delimit one's own place in a world bewitched by the invisible powers of the Other." It is an attitude typical of modern science, politics, and military strategy. In Mahan's writings, strategy is specified in opposition to tactics. The former addresses factors and principles that are applicable to all ages, whereas the latter addresses the historical and the contingent. In *The Influence of Seapower upon History* he writes:

> The considerations and principles which enter into [the growth of sea power] belong to the unchangeable, or unchanging order of things, remaining the same, in cause and effect, from age to age. They belong, as it were, to the Order of Nature, of whose stability so much is heard in our day; whereas tactics, using as its instruments the weapons made by man, shares in the change and progress of the race from generation to generation.

Tactics will inevitably change from historical period to historical period, but "the old foundations of strategy so far remain, as though laid upon a rock."[47]

De Certeau identifies three important effects of the strategic gaze. First, it involves a triumph of place over time. It is a mastery of time through the foundation of an autonomous place. This "autonomous place" in Mahan is the "Order of Nature." His spatialization of history is achieved not by any explicit transposition of biological laws onto human history but by the representation of geographical factors as material and permanent, part of the timeless Order of Nature, which exercises an enduring, stagelike influence on the immediate spectacle of history.[48]

Second, the strategic gaze is also a mastery of places through sight. "The division of space makes possible a *panoptic practice* proceeding from a place whence the eye can transform foreign forces into objects that can be observed and measured, and thus control and 'include' them within its scope of vision" (36; original emphasis). In Mahan's text, the strategist has a gaze that enables him to distinguish the historical from the eternal, the transitory from the permanent, the fluid from the rock. With this God's-eye gaze comes a proclivity to play at being a prophet. "To be able to see (far into the distance)," de Certeau notes, "is also to be able to predict, to run ahead of time by reading a space." Both the hubris one finds within the geopolitical tradition and also the sensationalism one finds in writings about the geopolitical tradition is a function of the arrogance promoted by the assumed exceptionalism of its insight. Indeed, this assumed exceptionalist vision, as we will explore in chapter 4, gave the theory and practice of geopolitics a quasi-religious tone in many cases.

Finally, de Certeau notes that not only is the strategic gaze a form of power that transforms the uncertainties of history into readable spaces but that institutional power is a precondition of its operation. Military or scientific strategies, he notes, have always been inaugurated through the constitution of their "own" areas, the "neutral" and "independent" institutions, laboratories, and institutes where strategic observers can perform their "disinterested" research. Such institutions, which is where one frequently finds research on "geopolitics," are literally seeing centers, panoptic towers that represent themselves as sovereign sites/sights from whence global politics can be surveyed and known with certainty.

It should also be noted that the position of the strategist is a highly gendered one that functions in an antigeographical manner not despite but because of its privileging of space over time. Privileged within Mahan's text, and Cartesian perspectivalism more generally, is a *disembodied* spectator subject who is above and beyond a *situated* space and time. This elevated, transcendent (Western) seeing Man can unproblematically gaze across the vast spaces of the earth's oceans and continents, and also across the many episodes of human history, from Greek and Roman times to the present. Geography is thus rendered spaceless and history timeless within the operation of this epistemological system in that both are taken to be transcendental coordinates of the universal nature of things.

Rudolf Kjellen (1864–1922) and the Linnaeanization of Global Space

Built upon the spatialization of history in the geopolitical tradition is, as we have already noted, a normative imperialist political agenda. Mackinder's writings are motivated by his interest in the British Empire reforming and restructuring itself so it could more effectively and efficiently address the social, economic, and transportational transformations of the early twentieth century. His constant preoccupation was the growth of German power on the European continent. Within the German Empire, the writings and imperialist philosophy of Friedrich Ratzel were inspiring a second generation of geopolitical theorists. Ironically, it was a Swedish political scientist, Rudolf Kjellen, who played a key role in congealing geographically sensitive commentary on international politics into what would later be recognized as a geopolitical approach to international politics in Germany after Ratzel's death in 1904. Kjellen was also the intellectual who first coined the term "geopolitics" in an article on the boundaries of Sweden in 1899.[49] His work found a more receptive audience in Germany than in his native Sweden, where he served for a time as a right-wing member of the Swedish Conservative Party. Kjellen's politics were part of the conservative reaction within Sweden to the Norwegian nationalist movement's push for an independent Norway in the 1890s. He opposed Norway's independence vociferously both in print and in parliament. After losing this battle (the Swedish-Norwegian union was dissolved in 1905), he supported the cause of the German Empire in Europe. His survey book *The Great Powers of the Present* went through twenty-two editions in German between 1914 and 1930, the last edited by the German geographer Karl Haushofer.[50] His interpretation of World War I as a conflict between the "ideas of 1789" (freedom, equality, and fraternity, represented by England and France) and the "ideas of 1914" (order, righteousness, and national solidarity, represented by Germany), with the latter inevitably triumphing because of the decadence of the former, was published in Germany under the title *Ideas of 1914* (1915) and proved immensely popular.[51] Kjellen's elaboration of Ratzel's biogeographical reasoning in *The State as an Organism* (1916) and *Foundations for a System of Politics* (1920) were well received by Haushofer and his colleagues who, after the war, would take Kjellen's concept of geopolitics and build it into a distinctly German school of geopolitical reasoning.

Kjellen's use of the concept of geopolitics must be understood as part of his ambition to provide political science with a Linnaean system to study the state. Like Ratzel, Mackinder, Mahan, and other intellectuals in the geopolitical tradition, Kjellen emphasized the "natural" and "organic" attributes of the state in counterdistinction to the legalistic and juridical understandings dominant at the time in political science.[52] *Geopolitik* was the first of five attributes of the state, and it involved the study of the territory of the state. The other attributes were *Demopolitik* (the population of the state), *Ekopolitik* (the economic structure of the state), *Sociopolitik* (social politics), and *Kratopolitik* (governmental-constitutional politics). *Geopolitik* was the most developed and most explicitly Ratzelian of Kjellen's categories, and he specified it as the study of the state as a "geographical organism or a phenomenon in space."[53] Kjellen further subdivided the category into the study of a state's relative location, the form of its territory, and its size. He distinguished between what he called first-rank "world powers" (England, Germany, Russia, and the United States) and second-rank "great powers" (Austria-Hungary, France, Italy, and Japan). Like Ratzel, he believed that the future lay with large autarkic continental imperialist states whose territory was compact and contiguous, with railways facilitating communications and the accumulation of power.[54]

Karl Haushofer (1869–1946) and the Spatialization of Imperialist Desire

Karl Haushofer took Kjellen's concept of geopolitics and transformed it into what he promoted as a whole approach to the study of geography and global politics. Rather than being merely one part of the study of the state, *Geopolitik* in Haushofer's writings became the name for a comprehensive spatialization of German desire to break from the strictures of the Versailles Treaty and become a great power once again. Haushofer was a Bavarian army officer and instructor in military history who served for two years as an artillery adviser in Japan between 1908 and 1910. Upon his return to Germany, he completed a Ph.D. in geography, geology, and history at the University of Munich addressing the German influence on the development of Japan. During World War I Haushofer served with the German army on both the eastern and western fronts, rising to the rank of major general by the end of the war. Haushofer began his academic career at the University of Munich in 1919 at the age of fifty. Among his admirers was Rudolf Hess, his aide-de-camp during the war and a subsequent student of political geography at the university. It was Hess who first introduced Haushofer to Adolf Hitler in 1922. During the time of Hitler's imprisonment in Landsberg, Haushofer gave him a copy of Ratzel's *Political Geography* while he was dictating *Mein Kampf* to Hess. Haushofer's influence on Hitler, as we will see in chapter 4, was the subject of a significant Allied propaganda literature during World War II, but it is now generally conceded that his influence on Hitler was wildly exaggerated.[55]

Socially conservative and elitist in his thinking, Haushofer was sympathetic to many of the stated aims of the Nazi Party without ever being a Nazi. In the 1920s Haushofer delivered public speeches to Nazi meetings and actively campaigned, like many others, against the Treaty of Versailles. Haushofer also kept up his contacts with the German army through Oskar von Niedermayer, writing secret reports for the army on Japan and the situation in East Asia in the 1920s. Haushofer was involved in the founding of two significant institutions in this decade. The first was the establishment of the German Academy, which was formally opened in May 1925. The aims of the organization, as written in its rules and regulations, were to "nourish all spiritual expressions of Germandom and to bring together and strengthen the unofficial cultural relations of Germany with areas abroad and of the Germans abroad with the homeland, in the service of all-German folk-consciousness."[56] Haushofer, who headed up the "Practical Department" of the Academy, became its president in 1934 in a move designed to appease the newly entrenched Nazi leaders of Germany. Slowly over the next five years the organization was taken over and made to serve Nazi ends with Haushofer's help. During the war the organization was placed under the auspices of the Propaganda Ministry. Although it was never a central spy organization, the academy seems to have been the inspiration for the myth of a "Geopolitical Institute" fostered by sensationalist stories in the Allied press after the outbreak of World War II.

The second was the establishment of the *Zeitschrift für Geopolitik* (Journal of Geopolitics) in 1924, with a number of collaborators. Haushofer's aim was to create a popular and influential journal that would address contemporary geopolitical issues and problems. Geopolitics, for Haushofer, was inseparable from practical politics; it was an aid to the conduct of statecraft. In 1928 he and the other editors of the *Zeitschrift* outlined their definition of geopolitics:

Geopolitics is the science of the conditioning of political processes by the earth. It is based on the broad foundations of geography, especially political geography, as the science of political space organisms and their structure. The essence of regions as comprehended from the geographical point of view provides the framework for geopolitics within which the course of political processes must proceed if they are to succeed in the long term. Though political leaders will occasionally reach beyond this frame, the earth dependency will always eventually exert its determining influence. As thus conceived, geopolitics aims to be equipment for political action and a guidepost in political life.... Geopolitics wants to and must become the geographical conscience of the state.[57]

Geopolitics, for Haushofer, was an objective science based on the study of natural phenomena and the laws of nature. Its methods were the methods of evolutionary biology and natural science, which, as we have already noted, were panoptic and spatializing in their conception. Geopolitics studied the state as an organism and the inevitable competition among state-organisms for living space. In conceptualizing political processes in this way, Haushofer perpetuated Ratzel's biologization of international relations.

The *Zeitschrift für Geopolitik* was to be an instrument for the education of Germany's leaders and also an educational tool for the inculcation of geopolitical habits of thought among the German masses. By the early 1930s the *Zeitschrift* had a circulation of 3000 to 4000 copies. The founding of an *Arbeitsgemeinschaft für Geopolitik* (geopolitics study group) in 1932 and its subsequent development and financing by the Nazis helped increase the circulation of the *Zeitschrift* to 7,500.[58] Haushofer later described his goal as the overcoming of a "lack" within Germany at the time. "What seemed most lacking in the resumption of the educational process for the training of German youth after the war was the capability to think in terms of wide space (in continents!) and the knowledge of the living conditions of others, namely oceanic peoples." Germany's problems, according to Haushofer, stemmed from its lack of a broadness of thought. Its thought was "limited by a continental narrowness as well as a smallness in its world vision."[59] Geopolitics sought to broaden the geographical vision of the German nation; it was about teaching the German people to think in terms of global spaces. Read symptomatically, geopolitics was a spatialization of the imperialist desires of the small community of militaristic males who felt Germany was castrated by the Versailles Treaty. Instilling geopolitical habits of thought within the population was the first step for Germany to overcome its "lack" and secure for itself the vast spaces that defined all truly great powers.

Up until Hess's flight to England in May 1941, an incident in which Haushofer's son Albrecht played a part (thus landing himself in jail for a short period; he was later murdered by the SS in 1945), Haushofer appears to have gone along with the Nazis as a means of realizing this desire. Like Hitler, Haushofer believed Germany should strive for lebensraum primarily in the east rather than overseas in Africa or elsewhere.[60] Following Mackinder's stress on the importance of the heartland, Haushofer argued for the construction of a continental block comprising Germany, Russia, and Japan as a counterweight to the sea-based British Empire, which, by Haushofer's reading, was vulnerable and in decline. Haushofer saw elements of his strategy realized in the Anticomintern Pact of 1936 and the Nazi-Soviet Pact of 1939, but it was frustrated by Operation Barbarossa. By this time Haushofer was over seventy and his personal influence on the conduct of Nazi foreign policy insignificant.

In a statement to Father Edmund Walsh, of the Office of the Chief Counsel of the United States investigating prominent Germans for possible prosecution in Nuremberg after the war, Haushofer offered his justification of German geopolitics. In a passage indicative of how biogeographical reasoning so pervaded his thought, Haushofer claimed that geopolitics was "legitimate" up until "the disturbance of its natural growth from 1933" by the Nazis.[61] If the Nazi takeover perverted a naturally growing geopolitics, Haushofer nevertheless approved of Nazi foreign policy until 1938, supporting also government on the basis of the *Führerprinzip*. He believed that Germany should have been satisfied with the outcome of the Munich conference (at which he was an adviser) and claimed that he made this clear to Hitler, with whom he fell into disfavor after a meeting in November 1938, their last meeting.[62] Haushofer claimed that German geopolitics from 1919 to 1932 had "goals quite similar to American geopolitics." Both wished to "achieve the possibility of excluding disorders in the future, like those of 1914 to 1918, through mutual understanding of peoples and their potentialities to develop on the basis of their cultural foundations and living space."[63]

A few months after writing this last will and testament of German geopolitics, Haushofer and his wife committed suicide on the grounds of his Bavarian estate. Walsh poignantly described how his "body lay sprawled on the ground, face down, his feet towards the creek, his hands clutching the Bavarian soil which he so passionately loved and so often described in his writings on *Lebensraum*."[64] Haushofer's attempt to draw an equivalence between German geopolitics and American geopolitics was a transparent effort to elide his Nazi sympathies and avoid responsibility for the murderous consequences of the Nazi drive for lebensraum. Nevertheless, his claim that what he taught corresponded with what was taught at the School of Foreign Service at Georgetown University (a school Walsh founded and where he taught geopolitics) or that a German manual of geopolitics "could as well have originated in the intellectuals and geopolitical institutions of any foreign country and of any of the allied nations" was not easily refuted. Haushofer's argument was prompted by Walsh's giving him the syllabus of his own course on geopolitics at the School of Foreign Service at Georgetown University. Gallagher, Walsh's biographer and fellow Jesuit, records Haushofer's reaction (without citation) to have been: "What complete material and means you had, and they complained about my so-called Institute of Geopolitics."[65]

Haushofer's reaction was understandable, for geopolitical thinking had made significant inroads into American political life by the end of the war. Walsh believed there was considerable scientific merit in the teachings of Haushofer, although he also believed these same teachings to be "an apology for international theft."[66] According to Peter Coogan, after returning from his work for the International Military Tribunal, Father Walsh "set out to transform the Georgetown School of Foreign Service into an American version of Haushofer's Institute of Geopolitics."[67] Whether this is the case or not, the School of Foreign Service was a vehicle for Walsh's lifelong crusade against world Communism, a crusade that began after Walsh's experiences as director of the Papal Relief Mission to Russia in 1922–23 (the school was renamed the Edmund A. Walsh School of Foreign Service in 1958). This mix of Jesuit religious crusade and geopolitical thinking congealed into an uncompromising anti-Communist geopolitics that exercised considerable influence over America's future diplomats and leaders. At the time of Walsh's death, for example, President Eisenhower recalled that "in 1928 I had the rare privilege of listening to a magnificent lecture of his on the growing menace of Communism. I think that I could recite some parts of its today."[68] According to David Halberstam, Father Walsh was the person who implanted the idea of a crusade against Communism in the mind of Joseph McCarthy at a dinner they had on January 7, 1950. Walsh's exposition on his favorite theme of Communism was picked up by McCarthy, who ran with the issue as a way to excite voters and get their attention.[69]

Other congealments of an American geopolitics, stimulated and influenced by contact with Haushofer's writings or, more accurately, made visible as "geopolitics" by the phenomenon of German geopolitics, could be found in the advocacy work of Lieutenant Colonel William S. Culberston (the dominant force within the short-lived Geopolitical Section in the Military Intelligence Service of the War Department), the logistics and planning lectures of West Point's Colonel Herman Beukema, the strategic writings of Edward Mead Earle and Harold and Margaret Sprout, the columns and books of journalist Walter Lippmann, the writings of Isaiah Bowman (examined in chapter 5), and also the writings of émigré professors of international relations like Arnold Wolfers and Nicholas John Spykman.[70] Although deploying often quite distinct institutional gazes, all were implicated in a general Cartesian perspectivalist enframing of the world, a producing of worldwide space as a global qua total geopolitical scene.

Nicholas J. Spykman (1893–1943) and the (Im)possibility of the Geopolitical Gaze

Nicholas Spykman was a Dutch emigrant to the United States who, like Ratzel, began his career as a foreign correspondent, although in his case in the Near East (1913–16), the Middle East (1916–19), and the Far East (1919–20). In 1923 he earned a doctorate from the University of California, writing his thesis and a subsequent book on Georg Simmel. He served as an instructor in political science and sociology at the University of California from 1923 to 1925 before going to Yale, where in 1935 he became chair of Yale's Department of International Relations and director of Yale's Institute of International Studies. Initially a Wilsonian idealist, Spykman was greatly influenced by Arnold Wolfers, who had fled his position as director of the respected Hochschule für Politik in Berlin when the Nazis came to power, joining the faculty at Yale in 1933. Wolfers converted Spykman to realist power politics and suggested that he immerse himself in the burgeoning field of geopolitics. With support from the Rockefeller Foundation, Spykman began research that would eventually result in the publication of *America's Strategy in World Politics* (1942) and *The Geography of the Peace* (1944), the latter a work published after Spykman's death from cancer at the age of forty-nine.[71]

Spykman is best remembered within accounts of the geopolitical tradition for his reworking of the Mackinderian emphasis on the "heartland" to stress the significance of the "rimland" areas of Eurasia, particularly Western Europe and Southeast Asia. Articulating an argument that was destined to become hegemonic within postwar U.S. strategic discourse, Spykman argued against the fallacies of American isolationism and the need for an active interventionist U.S. foreign policy to prevent any single power from dominating the Old World or Eurasian continent, a situation that would pose a grave threat to the security of the Western hemisphere.

In a 1938 two-part essay, "Geography and Foreign Policy," Spykman conceptualized geography in a manner similar to Mahan. Geography, Spykman declared, is "the most fundamental conditioning factor in the formulation of national policy because it is the most permanent. Ministers come and ministers go, even dictators die, but mountain ranges stand unperturbed."[72] This distinction between the order of the political-historical (ministers, dictators) and the order of the natural-geographical (mountain ranges) is, as we have already noted, a recurrent feature of the geopolitical gaze. But it is ironic that Spykman should use mountains as the signifier of the natural, permanent, and stable foundations of power when physical geography argued that mountains are not static objects but products of plate tectonics and dynamic cycles of erosion and deposition. This aporia in Spykman's reasoning is not insignificant, for Spykman's appeal is to a rhetorical and figural understanding of "mountains" as *signs* of permanence and durability.

This appeal to the rhetorical and the figural belies the foundational claim of Spykman's theorizing. This claim is that "geography does not argue; it just is." Geography is a permanent, self-evident realm of necessity that is present to itself; it is a durable, immanent force in international relations. Geography is independent of our beliefs and attitudes about it. It is part of the realm of necessity which is Nature. This is to be distinguished from the realm of freedom which is the world of Man. Geography is literally beneath the latter world, the world of the social, political, and ideological, the world of rhetoric and argumentation. It is a conditioning stage that makes demands on human history.

The problem with this argument is that it is blind to its own conditions of possibility. Geopolitics is a mode of theorizing that positions itself outside of the realm of both necessity and freedom. It scripts itself as a "view from nowhere."[73] But this "view from nowhere" is produced by and dependent upon its situatedness within Western thought. In scripting "geography" in the above manner, Spykman is already situating himself within discourse. In holding to the view that geography "just is," Spykman is already participating in the realm of rhetoric and argumentation. The faultline in the structure of the geopolitical gaze, the aporia in its division between the natural ("just is") and the political ("argue"), is that this very distinction must be transgressed for geopolitics to make sense. The condition of possibility of the geopolitical gaze is also its condition of impossibility. The geographical objects (for example, rivers, mountains, islands, continents, oceans), attributes (for example, size, natural resources, relative location, topography, climate), and patterns (for example, heartland-rimland, East-West, New World–Old World, continental-oceanic, land power–sea power) found in geopolitical scenes are not irreducibly transparent entities but socially constructed signs and systems of signification. Geopolitical productions invariably appeal to and rely upon the rhetorical dimensions of these objects, attributes, and patterns in order to persuade, to make their arguments convincing (as Spykman does with the notion of "mountain ranges"). The *visuality* of geopolitical productions, in other words, is dependent upon an unrecognized and unacknowledged *textuality*.

The claim that geography "just is," after all, is already an argument. It seeks to represent geography as a natural text and suppress the acts of interpretation necessary for it to be made meaningful. Geography, in the geopolitical tradition, is taken to be a finished, fully meaningful visual text with complete, monological messages of transcendent significance. But decoding these messages invariably involves geopolitics with questions of interpretation, textuality, and argumentation, questions that geopolitical productions attempt to contain and repress by Cartesian appeals to an external nature independent of our beliefs and knowledge about it. Spykman's declaration that "geography does not argue; it just is" works in this manner. But, the very identification and social representation of an "is" (the question of Being) is already an argument. Spykman's "argue" and "is" are not the opposites implied in his declaration. To claim an "is" is already to argue. The foundations of geopolitics are not rock solid and natural but fully social and inescapably political.

THE IRONY OF THE GEOPOLITICAL GAZE

The great irony of the geopolitical gaze (which, it should be stressed, is heterogeneous) as a congealed horizon of geo-power is that its functioning depends on a *suppression* of geography and politics. In its spatializations, biologization, linnaeanization, strategization, and naturalization of the historical, it works to degeographicalize and depoliticize the study of international politics. First, geopolitics is a form of geography that requires the systematic forgetting of the struggle over geography in order to make sense. Although it supposedly addresses the question of geographical difference in the world, the geopolitical gaze spatializes a world within a system of ethnocentric sameness. By gathering, codifying, and disciplining the heterogeneity of the world's geography into the categories of Western thought, a decidable, measured, and homogeneous world of geographical objects, attributes, and patterns is made visible, produced. The geopolitical gaze triangulates the world political map from a Western imperial vantage point, measures it using Western conceptual systems of identity/difference, and records it in order to bring it within the scope of Western imaginings. Geopolitics depluralizes the surface of the earth by organizing it into essential zones (middle strip, heartland, rimland, New World, Old World, Eurasia), identities (continental, oceanic), and perspectives (the seaman's point of view, the landsman's point of view). In sighting a world within the terms of Western forms of knowledge, geopolitics is siting it within the desires and fantasies of a privileged community of "wise men" and also citing it within the textual tropes and discursive contours of hegemonic centers of knowledge and learning.

Here also is a tradition that is eminently political and recognizes itself as such, yet rests on a depoliticization of geographical and political processes. Two dimensions of this depoliticization are particularly significant for they had an impact on how geopolitics was used politically to justify organized violence and murder by the state. The first is how the representation of the state as an organism effaces it as a political and social entity whose very form was the product of ongoing historical and geographical struggles over the definition of the nation and the legitimacy of state borders. Although the conceptualization of the state as an organism was hardly new,[74] the neo-Lamarckian reading produced by Ratzel and others gives it a biogeographical twist. The state is written as a competitive entity whose survival and growth is dependent upon its ability to adapt itself successfully to new technologies and skills so it can exploit its environment to the maximum extent. As we have already noted, this type of reasoning provided a justification for imperialist expansionism in Germany. In postwar South American geopolitics (a form of geopolitics that drew particular inspiration from German geopolitical writings), it developed into an ideology that emphasized the internal health of the state-as-organism and represented those pushing for social change or protesting military rule as dangerous "cancer cells" that needed to be eliminated for the health and prosperity of the gendered body politic.[75] It, in short, helped justify the bloody internal repression and state-sponsored murder that characterized the South American national-security state.

The second dimension is the depoliticization of certain political processes by representing them as inevitable and eternal processes of nature. Imperialism, territorial expansionism, and militarism are depoliticized in geopolitical texts as natural processes and not the highly contentious political, social, and economic processes they are. The outbreak of wars and conflicts is read as merely an instance of a perpetual struggle for survival and power among competing states. In this facet at least, geopolitics is a variant political realism, a variant that enlists geographical factors to write wars as inevitable. Mackinder, for example, explained the First World War within the terms of his timeless struggle between land power and sea power and between East and West. In the opening section to *Democratic Ideals and Reality,* the work he addressed to the leaders gathering at Versailles to redraw the map of Europe after the war, he declared, "The great wars of history—we have had a world war about every hundred years for the last four centuries—are the outcome, direct or indirect, of the unequal growth of nations, and that unequal growth is not wholly due to the greater genius and energy of some nations as compared to others; in large measure it is the result of the uneven distribution of fertility and strategic opportunity upon the face of the globe." There was, then, an innate logic to the geography of the earth that made wars more or less inevitable. There is in nature no such thing, he continues, as equality of opportunity for nations.[76] Nicholas Spykman offered his own version of this depoliticizing argument by contending that "because the geographical characteristics of states are relatively unchanging and unchangeable, the geographic demands of those states will remain the same for centuries." Since the world was a limited and confined space, these demands invariably caused friction. Thus, he concludes in a stark geodeterministic manner, "at the door of geography may be laid the blame for many age-long struggles which run persistently through history while governments and dynasties rise and fall."[77]

The ironic suppression of both geography and politics in the operation of geopolitics has its own geographical and political consequences. In using Eurocentric standards to survey, measure, and describe the world's geography and political structure, such reasoning is a form of epistemological imperialism. It functions as a geo-power that seeks to secure the claims of particular Western places and powers to write the history and geography of all. The geopolitical gaze, in other words, is itself profoundly geo-political; it seeks to enforce the vision of space and power of a certain metropolitan spatial and political order over those marginalized groups, either within metropolitan societies or within the colonies, who would contest that order. But contested that order would be.

2

Critical Geopolitics

> *A whole history remains to be written of* spaces *which would at the same time be the history of* powers *(both these terms in the plural) — from the great strategies of geo-politics to the little tactics of the habitat.*
> — MICHEL FOUCAULT[1]

One of the great ironies of the discipline of modern geography is that it remained remarkably blind to the politics of its own gaze and geographical history for so long. The discipline made vital contributions to the war effort in all states, yet, after the war, nearly all its leading figures avoided reflection on the politics of the discipline, particularly geopolitics, which was simply stigmatized because of its presumed association with the Nazis. Many of those who held positions of authority in the discipline in the postwar period—in the United States (like Richard Hartshorne, whom we will discuss in chapter 5), the United Kingdom,[2] and Germany[3]—worked in the various intelligence services during the war and did not, by and large, care to question or object to the political culture of the Cold War. It was not until the formal decolonization of the European empires beginning in the late fifties and the emergence of a small but significant dissident intellectual culture in the sixties that the imperial heritage of modern geography came to be recognized and questioned. Civil rights struggles, protests against the Vietnam War, and a growing urban crisis within the United States provoked the emergence of a self-consciously radical geography in the Anglo-American realm in the late sixties. Likewise, in France the political and intellectual ferment surrounding Vietnam, Algeria, and 1968 also produced a radicalized geography that began to dismantle the unquestioned assumptions and epistemological infrastructure of the modern discipline. The publication of Yves Lacoste's *La Géographie, ça sert, d'abord, à faire la guerre* (Geography is first and foremost about making war) in 1976 marked a new critical beginning in the historiography of the discipline.[4] Lacoste's work, examined in detail in chapter 5, was a polemical statement on the functioning of geography as a discipline that supported both an imperial order overseas and a capitalist order at home. Together with the works of others (who, like Lacoste, found an outlet for their work in *Antipode,* an important radical geography journal established in 1969), it helped politicize geographical knowledge and the study of space as socially produced. Most important, it pointed to the need for a thorough revisionism within the discipline of modern geography. For the first time, the challenge of decolonizing geographical knowledge and space could be conceptualized as such. The postmodern disorientation of modern geography had begun.[5]

Political geography within the Anglo-American realm was, however, in a somewhat moribund state during the early seventies. Henry Kissinger's revival of the term "geopolitics" gave it a new generic meaning as a synonym for balance-of-power politics on a global scale. This not only erased the Nazi stigma associated with the term but, more significantly for political geographers, foregrounded the problematic of geography and global politics anew. Few, however, addressed the specifics of Kissinger's articulation and recasting of this long-standing problematic. Political geography's predominant focus was not on the production of the global space but on national and municipal issues in electoral geography.

This changed somewhat in the early eighties with the establishment in 1982 of the journal *Political Geography Quarterly* (edited by Peter Taylor and John O'Loughlin), which built on the politicization of geographical discourse begun in the pages of *Antipode* and explicitly encouraged research at the international and global scales.[6] More significant, however, to the revival of interest in the broad question of geopolitics was the election of Ronald Reagan as president. Whereas Kissinger's geopolitics in the 1970s was about détente between the superpowers and controlled competition in the

Third World, the Reagan administration's "new geopolitics" was a return to explicit ideological competition between the superpowers as a means of reasserting the hegemony of the United States within the world.[7] Traditional U.S. spheres of influence like Central America and the Caribbean were the sites of a new hysterical politics of fear concerning Communism. Both regions were rezoned into a newly conceptualized "Caribbean basin" space in the U.S. geopolitical imagination, a space where the patriarchal authority of the United States was under challenge and demanded a tough "hardheaded" response.[8] The civil war in tiny El Salvador was conceptualized as posing a direct threat to the national security of the United States, as was political instability in the even smaller Caribbean island of Grenada. The Sandinistas of Nicaragua were also a dangerous threat in need of patriarchal punishment, a rabble of revolutionary upstarts who needed, in Reagan's words, to learn to say "Uncle" in dealing with the United States.

The scripting of the global scene by the former B-movie-actor president produced a geopolitics that was markedly televisual in its presentation and articulation. The cartographic supports Reagan employed in addresses to the nation featured patriotic blue space ("democratic" Central America) in conflict with foreign red space (Nicaragua and Cuba). The Soviet Union was scripted as the "Evil Empire" in a tale of conflict between the forces of light and the forces of darkness. This teletraditional revival of the Cold War as the dominant spectacle in international politics and rewriting of global space in the name of "geopolitics" provoked new interest in the functioning of the problematic of geopolitics.[9] Most of this research, however, did not reflect on the *genealogy* of geopolitics in any systematic way. Rather, the *term* "geopolitics" was studied within the specific historical conjuncture of its use by the Reagan administration and other pressure groups.

In the late eighties, John Agnew and I sought to address the concept of geopolitics in more comprehensive terms. In a paper eventually published in 1992, we began from the Foucauldian premise that geography as a discourse is a form of power/knowledge.[10] This led us to the claim that geopolitics "should be critically re-conceptualized as a discursive practice by which intellectuals of statecraft 'spatialize' international politics in such a way as to represent a 'world' characterized by particular types of places, peoples and dramas. In our understanding, the study of geopolitics is the study of the spatialization of international politics by core powers and hegemonic states" (192).

Implicit within our argument was an understanding of international politics as a spatial spectacle. The second Cold War was a remake of the original Cold War as B-movie farce. Our general concern was with the scripting of places in this spectacle, the spatialization of the stage of international politics. To this initial redefinition of geopolitics were added four theses. The first was the argument that "geopolitics is not a discrete and relatively contained activity confined only to a small group of 'wise men' who speak in the language of classical geopolitics." Rather:

> Simply to describe a foreign policy is to engage in geopolitics, for one is implicitly and tacitly normalizing a particular world. One could describe geopolitical reasoning as the creation of the backdrop or setting upon which "international politics" takes place but such would be a simple view. The creation of such a setting is itself part of world politics. This setting is more than simply a backdrop but an active component of the drama of world politics. To designate a place is not simply to define a location or setting. It is to open up a field of possible taxonomies and trigger a series of narratives, subjects and appropriate foreign policy responses. (194)

Our second thesis was the specification of two different types of geopolitical reasoning. Most geopolitical production in world politics, we suggested, is of a practical and not a formal type. Practical geopolitics refers to the spatializing practices of practitioners of statecraft such as statespersons, politicians, and military commanders. These intellectuals of statecraft are those who concern themselves with the everyday conduct of foreign policy. Formal geopolitics refers to the spatializing practices of strategic thinkers and public intellectuals who set themselves up as authorities on the totality of the world political map. These intellectuals of statecraft are employed in strategic institutes and other seeing and surveying centers (panoptic towers) within civil society. In contrast to the pragmatic, off-the-cuff reasoning of engaged statespersons, their reasoning is shaped by relatively formalized rules governing the production of spatializing descriptions, statements, and surveys of the international scene.

Our third thesis was that the study of geopolitical spatialization at the point of its production required consideration of the much broader question of the production of geographical knowledge within particular states and throughout the world system as a whole. In order to study the spatialization of global politics by intellectuals of statecraft, one needed to consider the embeddedness of these intellectuals within local, national, and transnational interpretative communities. To explain, for example, the Reagan administration's spatialization of "El Salvador" or "South Africa" one needed to explicate the importance of a local decision-making culture within the White House, the use made of the American mythology of national exceptionalism, and, finally, the significance of deeply embedded mores of Western thought concerning gender, race, and identity.[11]

Our final thesis placed the study of geopolitical reasoning within the context of the study of hegemony in a nonstatist, Gramscian-inspired sense. A hegemonic power like the United States is by definition a "rule writer" for the world community. Those occupying positions of power within the United States "become the deans of world politics, the administrators, regulators and geographers of international affairs. Their power is a power to constitute the terms of geopolitical world order, an ordering of international space which defines the central drama in international politics in particularistic ways" (195). What is important here is the activity of rule making and rule following rather than state dominance, for hegemony is more than the primus inter pares power of a state. One can have a condition of hegemony without a hegemonic state, although, in the period from 1945 to 1985, the rules governing world order were overwhelmingly shaped by the institutional power and disciplinary power/knowledge apparatuses headquartered in the United States.

This set of arguments marked the beginning of a broadly "postmodern" disorientation of (imperial) geopolitics, a dislocation of the concept so that it names a problematic much larger than its assumed meaning within the discourse of international politics. Rather than taking geopolitics for granted as either the name of a particular tradition of thinking about international politics or as part of the self-evident reality of international politics, we sought to reconceptualize it as the problematic of the social inscription of global space by intellectuals of statecraft. Commonsense conceptions of "geopolitics" in the 1980s are made the point of entry into the problematic of geo-politics, the politics of spatializing global politics. Orthodox conceptions of "geopolitics" are turned against themselves to reveal an unquestioned and unproblematized geographical politics at work in the scripting of the dramas of the global political stage. International politics has a *geo-politics* that is much more pervasive and ideologically significant than orthodox understandings of geopolitics would indicate. There is a geo-politics beyond, beneath, and within "geopolitics," a geopolitical gaze, a particular congealment of geo-power, which exceeds that which is normally conceptualized as the geopolitical tradition.

In his study of the geopolitical reasoning of certain intellectuals in the Committee on the Present Danger, an important late-seventies pressure group that did much to shape the first Reagan administration's scripting of the global scene at that time, Simon Dalby addresses these issues and themes, arguing, in his conclusion, for the creation of a critical geopolitics (his work is considered in detail in chapter 5).[12] In keeping with dissident international relations' concern to challenge the statist organization of global political space, Dalby argues that a critical geopolitics can challenge the geographing of (global political) space as a system of pregiven containers for politics, a state-centric system of space. It can "investigate how the categorizations and cultural creations through which we come to understand and write in turn shape our political existence" (173). The focus of such a critical geopolitics should be on "exposing the plays of power of grand geopolitical schemes, and in turn, challenging the categorizations of discourses of power" (180).

Such declarations conceptualized critical geopolitics in an abstract and general way. In a subsequent review article Dalby eloquently articulates the need for a critical geopolitics:

> What is being argued for here is nothing less than a recognition of the importance of studying the political operation of forms of geographical understandings, recognizing that geographs are specifications of political reality that have political effect. To construct critical political geographies is to argue that we must not limit our attention to a study of the geography of politics within pregiven, taken-for-granted, commonsense spaces, but investigate the politics of the geographical specification of politics. That is to practice critical geopolitics.[13]

Since then Dalby's plea for a critical geopolitics has been answered by a wide variety of literatures within the discipline of geography that seek, in multiplicities of ways, to investigate "the politics of the geographical specification of politics."[14]

But just what is meant by "the politics of the geographical specification of politics"? In this and other articulations, critical geopolitics seems to rest on general declarations about space and power. Despite the wealth of recent literature on critical geopolitics, there remains a pressing need to deepen and sharpen its character as a distinct intellectual and political project. Part of the reason why critical geopolitics has not yet distinguished itself in a sufficiently precise manner is its surprising failure to rigorously conceptualize and theorize the very object that supposedly defines it: geopolitics. As a concept, geopolitics is regularly evoked and knowingly used yet rarely problematized. There is not yet an adequate theoretical discussion of the functioning of geopolitics as a sign within critical geopolitics. Nor has there been an explicit theoretical discussion of how critical geopolitics should function and how it needs to engage what we have already identified as the geopolitical gaze.

As a means of moving critical geopolitics beyond its current superficial generalities to a deeper level of intellectual engagement with the problematic it marks, I propose to use Derridean deconstructionism to reflect theoretically on the following three questions: (1) the meaning of geopolitics; (2) the purpose of critical geopolitics; and (3) the problem of the geopolitical gaze. In so doing, I wish to elaborate some "methodological" principles that I believe can be used to deepen critical geopolitics as an approach.[15] These "methodological" principles inform the chapters that follow in a silent, stealthy manner, for my concern, after this chapter, will be to examine particular congealments of geopolitics in their con-textual detail.

THE TEXTUALITY OF GEOPOLITICS

One of the most common problems preventing the rigorous theorization of geopolitics is that the term itself is a polysemic one that has long exceeded its original con-textual specification by Kjellen, Haushofer, or even Kissinger. A cursory browse through any of the cites generated by a database search using the term reveals a wide variety of usages and meanings. There are, for example, references to the geopolitics of capitalism, environmentalism, race, urban zoning politics, and cinema, among others.[16] The qualifier "geopolitical" has an even greater range of usage.

This confusion over the meaning of geopolitics, while undoubtedly more intense today than before, is hardly new. From its emergence, the term "geopolitics" has had shifting and unstable meanings. Rudolf Kjellen, for example, complained that the Germans were misusing the very word he himself coined.[17] In 1938 the term had at least five different meanings, which led one commentator to conclude that it was therefore not surprising that "a certain amount of obscurity has accompanied its use."[18] Another commentator in 1954 suggested that the term has a wide variety of uses and, in some instances, appears to have no meaning at all.[19] In 1986 yet another commentator was moved to write that the difficulty with geopolitics is that "it is conceptually so broad that it can and does mean all things to all people."[20]

The concept of a "geopolitical tradition" is also dogged by the problem of meaning. No real consensus exists on how to specify this tradition in a definitive manner, for no one is quite sure about what is and what is not part of the tradition. Some codifications trace it back to antiquity and Aristotle, while others locate its origins in the late nineteenth century.[21] Some codifications postulate a tradition of "Western geopolitical thought" and confine its range to the work of geographers and selected nongeographers, while others place it within the tradition of Western strategy.[22] For some the tradition is composed of "wise men" and strategic thinkers only and not practitioners of geopolitics. Recently, there is a tendency to divide the tradition into a "classical" and "modern" geopolitics, although there is no consensus on how one marks the divide. Zoppo and Zorgbibe, for example, use the concept of "classical geopolitics" in order to foreground the rupture in global politics marked by the post–World War II spread of nuclear weapons, the conquest of space by military technology, and the rise of a "nuclear geopolitics" where the "cognitive space" of strategic decision-making is all important.[23] Sloan, however, uses the concept of "classical geopolitical theory" in order to demarcate a "good" Anglo-American geopolitics (Mahan, Mackinder, and Spykman) from a "bad" geographically deterministic German geopolitics.[24]

Given all this apparent confusion, it is tempting to simply declare "geopolitics" a meaningless term and the "geopolitical tradition" an imprecise notion. This, however, would be a mistake, for these very judgments are organized within the implicit horizon of a teleological theory of meaning. To declare something meaningless or imprecise is to assume to know the final state of full meaning and the precise use of concepts that are imagined to be stable and homogeneous. The attitude behind these assumptions is what Jacques Derrida has termed the logocentrism of Western thought, the dependence of theories of thought, discourse, or, in this case, meaning on a metaphysical authority (*Logos*) that is considered external to them and whose truth and validity they express.

As a means of challenging the operation of logocentrism, Derrida asserts the irreducible textuality of all concepts and terms. Terms and concepts do not mean anything in and of themselves. All concepts are produced within discursive networks of difference and are therefore dependent upon these networks of difference or infrastructures for their identity. The "teleological value of the homogeneity of concepts," Rodolphe Gasché explains, "is disproved by the very process of the formation of concepts."[25] He elaborates this Derridean argument by making the following four points. First, "since a concept is not a simple point but a structure of predicates clustered around one central predicate, the determining predicate is itself conditioned by the backdrop of the others" (128–29). Geopolitics, for example, is conditioned by the predicates "geography" and "politics," which are themselves delimited and conditioned by other predicates in unstable and indeterminate ways. Second, "each concept is part of a conceptual binary opposition in which each term is believed to be exterior to the other. Yet the interval that separates each from its opposite and from what it is not also makes each concept what it is. A concept is thus constituted by an interval, by its difference from another concept" (129). The identity and meaning of geopolitics, in other words, derives as much from that which is asserted and silently assumed to be not-geopolitics as from any positive value it claims as its own. All concepts are in a sense paradoxical. Their conditions of possibility, as I argued in chapter 1 about geopolitics, are also their conditions of impossibility. Third, "concepts are always (by right and in fact) inscribed within systems or conceptual chains in which they constantly relate to a plurality of other concepts and conceptual oppositions from which they receive their meaning by virtue of the differential play of sense constitution, and which thus affect them in their very core" (129). Geopolitics, in other words, cannot be abstracted from the textuality of its use. Geopolitics can only be studied in terms of its embeddedness in the (general) text. Finally, "one single concept may be subject to different functions within a text or corpus of texts" (129). Geopolitics is an exemplary case in point, for it can refer to either a theory or a practice, a concept or "reality," an aspect of international politics or a summation of its totality.

Gasché's fundamental point, and that of Derrida, is that concepts are paradoxical entities or contradictory unities that can only be studied in their general textuality. A deconstructionist approach, therefore, does not directly answer the question "What is [geopolitics]?" This question, Derrida notes, is the instituting question of (Western) philosophy, a question he associates with this tradition's need to find an essence or nucleus of meaning in some transcendental signified.[26] Rather than submit to the rule of this question, which anticipates an essential answer, Derrida's strategy is to displace the question, to problematize its limits and conditions of possibility. Rather than work within the metaphysics of presence, he investigates the play of conceptual meanings in the reading and writing of (worldly) texts. Instead of assuming that geopolitics or the geopolitical tradition have self-evident and self-present identities, deconstructionists ask how "geopolitics" and the "geopolitical tradition" have been textualized with certain meanings at various times and in various contexts. How, for example, has the term "geopolitics" been charged with particular meanings and strategic uses within differing networks of power/knowledge? How has it been put to use in differing times and places? The term "geopolitics" poses a question to us every time it is knowingly evoked and used.

There are two important implications of these arguments for the practice of critical geopolitics. The first is that we should remain sensitive to the *heterogeneous histories* of "geopolitics" as a twentieth-century concept that functions as a gathering point for the production of geographical meaning about "international politics." Many of these histories have been forgotten by students of international relations. The second is that the investigation of geopolitics, nevertheless, exceeds the nominal functioning of the term "geopolitics," for this very term itself is dependent upon an infrastructural unthought that deserves our analysis. A critical geopolitics should not be driven by a nominal obsession with the term "geopolitics" but should strive to address the problematic of *geo-politics*, the general problematic of the scripting of global space by state-society intellectuals and institutions. It is a problematic that exceeds "geopolitics" as we know it or, more accurately, as we think we know it. This particular problematic, as I have already suggested, can in turn be considered as part of a larger problematic of geo-power, which concerns the mutually defining interdigitation of geography and governmentality in modernity.

PROBLEMATIZING AND DISPLACING GEO-POLITICS

In hyphenating geopolitics in the above manner I seek to place the putative stability and unity of geopolitics in question. It is a means of problematizing geopolitics, a way of putting it under erasure so the hidden logocentric infrastructures that make it possible are exposed to view. Geo-politics does not mark a fixed presence but an unstable and indeterminate problematic; it is not an "is" but a question. The hyphen ruptures the givenness of geopolitics and opens up the seal of the bonding of the "geo" and "politics" to critical thought. In undoing the symbolic functioning of the sign, its semantic instability, ambiguity, and indeterminacy are released. The sign lies open before us, a disrupted unity in question, a sign of a textual weave involving geography and politics.

This tactic of hyphenation follows a Derridean tactic of play whereby an old name is retained and subverted to designate the previously invisible problematic of its own functioning.[27] Geo-politics is the other of geopolitics; it is the unnamed logocentric practices that make it possible. Official articulations of geopolitics attempt to dissimulate the geo-politics that makes them possible, to repress their conditions of (im)possibility. The logic of using a disruptive play on the name of the old/official to designate an entirely new field of problematizations is historical and practical. Geo-politics communicates with geopolitics; it recalls its history yet destabilizes this history, both placing our inquiry within a certain realm of meaning yet also displacing the ways this realm of meaning delimits itself. The hyphenation is a lever of intervention that maintains a grasp on the previous organization yet also effectively transforms it.[28] It is a tactic that can be used occasionally (and sparingly) to simultaneously problematize and subvert. It should not, however, be used as the basis of an essential distinction between geopolitics and geopolitics. Geopolitics and geo-politics are one. Geopolitics, as I have already noted, is permanently hyphenatable.

Critical geopolitics is distinguished by its problematization of the logocentric infrastructures that make "geopolitics" or any spatialization of the global political scene possible. It problematizes the "is" of "geography" and "geopolitics," their status as self-evident, natural, foundational, and eminently knowable realities. It questions how "geography" and "geopolitics" as signs have been put to work in global politics in the twentieth century and how they have supervised the production of visions of the global political scene. Rather than innocent sites of declarative facts and constative statements about the world, these signs marks the site of space/power/knowledge production systems, operations that script the actors, settings, and dramas of global politics in deeply geo-politicized ways.

Critical geopolitics should not be understood as a general theory of geopolitics or an authoritative intellectual negation of it. As an approach that seeks to assert the irreducible textuality of "geography" and "geopolitics," critical geopolitics does not lend itself to the constative form; it is not an "is" but, in the manner of deconstruction, it *takes place*.[29] It is parasitical on that which it addresses, working within the con-textuality it explores to displace the infrastructures of geopolitics.

This taking place (as opposed to being-in-place) that is a displacement can be understood in a military sense. In contrast to the strategic ambition of imperial geopolitics (which is about the establishment of place or proper locus), critical geopolitics is a tactical form of knowledge. It works within the conceptual infrastructures that make the geopolitical tradition possible and borrows from it the resources necessary for its deconstruction. Geography and geopolitics are displaced into deconstructing themselves. Unlike the strategic mode of procedure, a tactical approach does not, de Certeau notes, delimit an exteriority:

> The space of a tactic is the space of the other. Thus it must play on and with a terrain imposed on it and organized by the law of a foreign power. It does not have the means to *keep to itself*, at a distance, in a position of withdrawal, foresight, and self-collection: it is a maneuver "within the enemy's field of vision," as von Bulow puts it, and within enemy territory.[30]

As an active intervention and commentary on the global political scene, critical geopolitics does not offer the strategic surveyor's perspective of the foreign policy "expert," with its reduction of time to space and detached, authoritative overview of a space it makes global, visible, and objectifiable. Rather, its mode of operation is a mobile, guerrilla one that uses what is at hand within a terrain governed by hegemonic political understandings in order to advance critical positions in a permanent war of position.[31]

THE GEOPOLITICAL GAZE:
PROBLEMATIZING CARTESIAN PERSPECTIVALISM

This contrast between tactical and strategic knowledge points to the need for critical geopolitics to more fully confront the ocularcentrism of the different condensations of geopolitics found throughout the twentieth century. Ocularcentrism, the domination of Western thought by the metaphorics of vision, has deep roots in the history of Western philosophy. Ancient Greek culture and philosophy privileged the faculty of sight as the noblest of the senses, a privilege that remains within the language we have inherited. As many have noted, the word "theory" has its root in *theoria*, which, in classical times, was the name for certain individuals (singular *theoros*), chosen on the basis of their general social standing in the polity, who were charged with the official task of observing the occurrence of some event and verbally certifying that it had taken place. This see-and-tell act, designated *theorein*, was a politically organized, public act of looking and surveying and not a private act carried out by an ordinary person.[32] The official act of beholding was politically powerful in that what was certified as having been seen could then become the object of public discourse. While this classical meaning does not determine the modern meaning of theory, it is nevertheless worth recalling as a salutary reminder of the interdigitation of seeing and surveying with systems of authority. Other modern words also reveal the pervasive ocularcentrism of Western thought. The Greek word *idea* means the look of a thing and comes from the same root as the Latin *video* (I see).[33] The word "evidence" also has its roots in *videre*, the Latin verb "to see," while "intelligence" comes from *intelligere*, "to see into, to perceive."[34] Like theory, the word "theater" also traces its roots to *theoria*, as does the word "theorem," which, as Jay notes, has allowed many commentators to emphasize the privileging of vision in Greek mathematics, with its geometric emphasis.[35] The metaphor of the stage, so crucial to the geopolitical gaze, was decisively shaped by the ocularcentrism of a Greek culture that conceptualized their gods as spectators and celebrated visual spectacles like theater.

Just as this ocularcentrism was significant in the Greek inventions of geometry and philosophy, so it was also crucial in the development of the Greek science of geography. Sight is preeminently the sense of simultaneity. It is intrinsically less temporal than the other senses and has thus long been associated with intellectual pursuits that tend to elevate static Being over dynamic Becoming, fixed essences over ephemeral appearances.[36] An epistemology structured by vision tends to configure knowledge in terms of the simultaneous display and full apprehension of all the elements of a given configuration. Simultaneity privileges space, conceptualized as a synchronic order of coexistences. Time is flattened and concentrated into a homogeneous synchronic itself.[37] Ocularcentrism does not simply privilege space at the expense of time; it produces both space and time as synchronic, not diachronic. Associated with this two-dimensionalization of space/time is a teleological drive toward totality. Greek geography, geometry, and cartography are all suffused with the teleological dream of displaying space as a simultaneous, synchronic totality. For Heidegger, these were general (Platonic) tendencies that reached full expression in a modernity that conquered, enframed, and (technologically) produced the world as a picture.[38]

With the privileging of the faculty of sight in Greek culture comes a denigration of language, writing, and rhetoric. Our modern prejudice against rhetoric can be traced to the classical Greek world, where the Sophists were maligned and language was considered inferior to sight. Rhetoric was banished from the realm of philosophy; it was the resort of the dishonest and the deceitful. What was initiated in Greek philosophy was augmented in the later history of Western thought by the innovations of perspectivalism and Cartesianism. Perspectivalist vision made a single sovereign eye the center of the visible world. Descartes elaborated how this eye was an inner eye of the intellect; it is the mind's eye and not the corporal eye that "sees." Descartes, according to Jay, tacitly adopted the position of a perspectivalist painter using a camera obscura to reproduce the observed world. This leads Jay to propose "Cartesian perspectivalism" as a shorthand term to characterize the dominant scopic regime of the modern era.[39] Cartesian perspectivalism reinforced the differentiation of the visual (sight) from the textual (cite), as Descartes assumed a divine congruence between language and the world of transparent objects (sites). The transcendent "sight" of the subject and its ability to discern inert "sites" was ultimately dependent upon the successful and sustained suppression of "cites" or signification. Sustaining such ocularcentrism, however, is not possible, for the problematic of signification and writing are already present in the very acts of seeing and siting. In exposing that which makes seeing possible and sites visible, a deconstructing critical geopolitics needs to demonstrate how the ocularcentric world of "sight" is a world that is already infested with textual "cites." In studying totalizing, detemporalizing spatializations of global(ized) politics, critical geopolitics must problematize the relationship between subject, object, and text, or, more prosaically, that between sight, sites, and cites.

Such intellectual critique is best undertaken concretely in the examination of specific strategic thinkers and worldly conjunctures, for Jay's notion of Cartesian perspectivalism is somewhat loose in its conception. While it is undoubtedly useful to speak of a dominant scopic regime in modernity, the particularistic operation of that regime varied considerably and was, as he himself points out, not without challenge, complications, or moments of unease.[40] Indeed, from the late nineteenth century onward, the dominance of Cartesian perspectivalism was under constant challenge from technological innovations, artistic movements, and philosophical discourse. Modern geography, however, appears to have been a force reconsolidating the dominance of Cartesian perspectivalism. Expressions of the meaning of geography frequently described it as the science of sight, simultaneity, and space. Under Paul Vidal de La Blache's influence, for example, the production of richly textured regional tableaux was considered the highest achievement of French geography.[41] Derek Gregory, as we have already noted, suggests that modern geography can be interpreted as a constellation of space/power/knowledge that sought to produce the "world-as-exhibition."[42] Certainly methodological discussions within both geography and geopolitics recurrently privilege the faculty of "visualization," the "art of seeing," and the unique "global view" of the geographer qua geopolitician. But such expressions of general ocularcentrism and certain commitments to Cartesian perspectivalism are best examined in their material specificity, for it is there we can best trace the particular aporias and blind spots that characterize the writing of geographical and geopolitical vision.

It should be noted that the problematization of Cartesian perspectivalism is not meant as a repudiation of ocularcentrism in toto. The metaphorics of vision infuse our inherited language and conceptualizations, and it would be folly to assume we could ever fully break from them. Critical geopolitics does have a perspective and a metaphorics of vision (it is, after all, a geopolitics), but it is a perspective and language of seeing dictated by that which it tactically engages. As de Certeau notes, the tactical is a maneuver within the enemy's field of vision. Its visions, therefore, are visions that seek to put vision in question; its seeing, a seeing that tries to reveal the unseen of seeing; its displays, dissident playings with practices of displaying; its insight, the insight that comes from the investigation of the infrastructure of insight. Its partial perspective is not that of the fully unified subject but that of the self-reflecting subject-in-formation, its theory not the detached and certified see-and-tell of the *theoria* but an interventionist traveling theory — theory resisting the ambitions of universal theory — that seeks to put the geo-politics of *theorein* in question.

The essays that follow are a series of tactical interventions into select aspects of the problematic of writing global political space. They are neither grand theory nor exhaustive survey but challenges to attempts to congeal meaning around geopolitics and discipline the open-ended textuality of global space. Before moving to consider them, let me add a final point about the concept of context, for if I have a grand and indeed insistent claim it is that geopolitics in all its forms — whether subdivided into formal, practical, and popular culture variants or not[43] — is best studied in its messy historical con-textuality. The notion of context, however, should not be commanded by a dominant referent or transcendental signified. Context is not a pure original point, an objective space/time coordinate, or a final resting place.[44] Context is an open structure, the limits of which are never absolutely determinable or saturated. Derrida has demonstrated how "writing" puts our conventional concept of context into question because it breaks with the original scene of intersubjective communication of conscious copresent subjects. Writing is about absence, deferral, and delay, about communication in the absence of the addressee. It must be iterable so that a text can continue to be read after its author has disappeared or died. Writing, therefore, exceeds our conventional concept of context; a text can be quoted in many different contexts and enter into new contexts and take on new meanings and significance. The writings of Alfred Mahan, for example, were quoted to justify and legitimate not only American naval expansionism but also naval expansionism in the contexts of imperial Germany and Great Britain. Even more remarkable, his texts were subsequently quoted by certain elements within the Japanese navy to make the case for an attack on Pearl Harbor, a location Mahan had described as not likely to be attacked. Similarly, the texts of Halford Mackinder and Nicholas Spykman were used to justify American Cold War militarism and particularly aggressive nuclear war-fighting doctrines in the 1970s and 1980s.[45]

Our notion of context, therefore, must recognize that there is no pure original context or scene of communication. Context is not a structure of presence but a structure of general textuality. The question of context, for Derrida, is a question of text that is not simply a written book or volume but a way of describing the inevitably signified or written nature of our social life. Derrida's famous claim that "there is nothing outside the text"[46] is at one with his argument about the concept of context: "Nothing exists outside a context."[47] His work reminds us of the "text" in "context" without dissolving the specificity of either.

Having said this, we would do well to avoid the narrow textuality of the more literary uses of Derridean deconstructionism, a conception of textuality that reduces questions of state power, technological development, and social structure to questions of literary ideology and discourse.[48] In critically investigating the textuality of geopolitics, we are engaging not only geopolitical texts but also the historical, geographical, technological, and sociological contexts within which these texts arise and gain social meaning and persuasive force. At the broadest historical level we are engaging, as Michael Mann has brilliantly described for the Great Powers in the period 1760 to 1914, the messy histories of the entwined development of capitalism and the state system, the rise of modern militarism, the growth of state infrastructural power, the tightening of state-society relations, the caging of social relations and class conflicts within "national" boundaries, and the resultant geo-politicization of greater and greater aspects of everyday social life.[49] That which congealed as "geopolitics" in the twentieth century is one of many points of entry into these macrohistorical and macrosociological processes, processes that are central to modernity as an enframed order of space and power.

3

Imperial Incitement

Halford Mackinder, the British Empire, and the Writing of Geographical Sight

> *Our aim must be to make our whole people think Imperially — think that is to say in spaces that are world wide — and to this end our geographical teaching should be directed.* — HALFORD MACKINDER[1]

Just after noon on September 13, 1899, Halford Mackinder rose to the peak of Mount Kenya along with his guide, Cesar Ollier, and the latter's porter, Joseph Brocherel. The moment was the climax of a Royal Geographical Society supported expedition organized by Mackinder and his friend Campbell Hausberg to conquer what he described in his journal as the "cold feminine beauty" of Mount Kenya, in a country officially known as "British East Africa." Not everything had gone smoothly with this conquest that took some 6 white European men and 170 natives across a land blighted by famine and smallpox. Mackinder and his party were attacked, supplies were sometimes scarce, and several of his porters were murdered.[2] Mackinder and his companions were the first Europeans to have conquered Mount Kenya, and it would be another thirty years before their feat was equaled. As Mackinder stood on the peak he carried out a plane-table survey of the mountain's upper part and marked off six valleys, inscribing them with names that included "Mackinder Valley" and "Hausberg Valley." After about forty minutes, with a storm threatening and ill-equipped for a stay, Mackinder and his companions returned over a hanging glacier, down the treacherous face of an arête to their base camp, "hungry and weary, but triumphant" (see Figures 3 and 4).[3]

The ascent of Mount Kenya by Halford Mackinder in 1899 was a landmark event in the career of the man best known today as one of the "founding fathers" of geopolitics. Born in England in 1861, Halford Mackinder was educated at Oxford, where he studied physical science, history, and law. As an extension lecturer in the 1880s Mackinder became a leading advocate of geographical education in Britain and played a key role in the eventual establishment of geography as a modern university discipline. Originally a liberal imperialist, Mackinder subsequently became an opponent of free trade and a strong advocate of tariff reform to unite the British Empire in the face of the growth of German power on the European continent. In 1910, he entered Parliament as a Conservative Unionist and served in the House of Commons until his electoral defeat in 1922. Subsequently he devoted his time to imperialist issues, serving as chair of both the Imperial Shipping Committee (established in 1920) and the Imperial Economic Committee (1925–31) (see Figure 5). During World War II his early ideas on political geography were hailed as prophetic and were organized under the sign "geopolitics," a word Mackinder never himself used and apparently disliked.[4] Mackinder died in March 1947, a week before the articulation of the Truman Doctrine and just as some within the Atlanticist foreign policy community were beginning to adapt his vocabulary on the "heartland" to explain why the Soviet Union was a threat that need to be contained.

Mackinder's desire to conquer Mount Kenya was a deliberate career move by a man seeking authority within the new discipline of geography in late Victorian Britain. At the end of the nineteenth century the so-called science of geography was, in the words of Joseph Conrad, "still militant but already conscious of its approaching end with the death of the last great explorer."[5] This approaching end was marked by the disappearance of "the exciting spaces of white paper" on European maps of Africa that Conrad had found so fascinating as a child. Such virgin white spaces were, of course, only blank to European eyes, the void marking territory not yet subjugated to European ways of writing and dominating the earth. The imperialism that adhered to this desire to write on the blank page, to penetrate and map the vast interiors of Africa, was a nineteenth-century

Figure 3. Man amid nature. Mackinder (?) next to giant lobelias, British East Africa, 1899. By kind permission of the School of Geography, Oxford University.

Figure 4. The object of conquest. Kenya peak and Teleki Valley from the southwest, British East Africa, 1899. By kind permission of the School of Geography at Oxford University.

Figure 5. Imperial man. Mackinder at the age of seventy-two. The original, by Sir William Rothenstein, hangs in the London School of Economics. By kind assistance of the School of Geography, Oxford University.

phase of a long-standing history.[6] From the fifteenth century, European knowledge-making apparatuses had been constructing the globe first in navigational terms.[7] The circumnavigation of the globe and the mapping of coastlines created a new planetary consciousness among literate Europeans. In the eighteenth century, these global(izing) projects were complemented by the emergence of one that took as its mission the systematization of nature and the production of detailed inventories and classificatory schemes for the mapping of continental interiors. This latter grid of knowledge was the natural history of Adanson, Linnaeus, and Buffon, an intellectual project markedly more ambitious than those that preceded it. The vast contents of continental interiors "would be known not through slender lines on blank paper, but through verbal representations in turn summed up in nomenclatures, or through labeled grids into which entities would be placed. The finite totality of these representations or categories constituted a 'mapping' not just of coastlines or rivers, but of every visible square, or every cubic inch of the earth's surface."[8]

Michél Foucault has traced the surface of emergence of this new realm of "natural history" in the classical age. For natural history to appear, he argues, it was necessary for history to become natural.[9] Histories, or discourses (in the form of books) on the meanings and uses of things, gave way to History, an order of knowledge with strict rules mandating a particular type of observation and documentation, separated from fable and hearsay. A somewhat similar logic appears to have operated in the case of the concept of geography, the details of which need research. The term's etymology suggested it as an "earth writing," a graphing of the geo. Geography in the sixteenth century referred to written texts, treatises, and discourses full of descriptions of the earth, accounts of classical learning and speculations that were as much theological as scientific.[10] It was a record of the divinely written "book of nature," of the system of signatures written into the being of things in the world.[11] By the seventeenth century, the understanding of geography as part of the prose of the world was challenged by a representational *episteme* that dissolved the profound kinship of language and the world. This was part of a large shift from reading the world as an intelligible text to looking at it as an observable but meaningless object.[12] A narrowly textualist understanding of "geography" gave way to an understanding of "geography" as the universal nondiscursive surface of nature. Where previously there had been many geographies, now there was only one definitive Geography, a complex inventory of physical and human objects that was slowly being assembled in the grid of European science.

A condition of possibility of this epistemic change, for Foucault, "is the space opened up in representation by an analysis which is anticipating the possibility of naming; it is the possibility of *seeing* what one will be able to *say*, but what one could not say subsequently, or see at a distance, if things and words, distinct from one another, did not, from the very first, communicate in a representation."[13] Natural history was enabled by the assumption that, on the one hand, words and things were separate yet, on the other hand, words could name things in a precise, neutral, and objective manner. Crucial to this activity was the faculty of sight, which became a central register of this regime of knowledge, while other faculties (touch, smell, taste) were relegated to the margins. Language, Foucault argues, had to be brought as close as possible to the observing gaze, and the things observed as close as possible to words. "Natural history is nothing more than the nomination of the visible."[14] It was a new field of visibility.

The representational scientific gaze that developed within natural history from the mid–seventeenth century challenged the tradition of reading and writing the earth in terms of divine and providential marks, although natural theology remained an important force in narrating the earth until well into the nineteenth century.[15] By this time, the figure of "Man" had emerged in Western epistemology. "Man" took the place of the king and God in modern thought.[16] He became both the object and subject of knowledge, an objective, transcendental recorder of things in the world while also being a subjective, empirical entity among these things. "Man," for those who wrote Geography, was "seeing-man," a transcendental European subject who was empowered with the sovereign power to see the world in the fullness of its positivity.[17] The eye of this seeing subject was sovereign, and it sighted nature as an inventory of sites. In her study of the narratives generated by European travel writers in the eighteenth and nineteenth centuries, Mary Louise Pratt notes how the lettered, male, European eye of these travel writers could familiarize and naturalize new sites/sights immediately upon contact by incorporating them into the nomenclatures and global grids of European forms of knowledge.[18] Sight sited geography by its power of citation. Only European seeing-man had the power to summon places before his eyes and write them up as sites. The eye was the seat of a monarchical author-ity to mark the land, to officially recognize (sight/

site/cite) its features and trope its territories into the grid of European learning.

To the British Foreign Office, the Royal Geographical Society (RGS), and aspirant professional geographers, the interior of "British East Africa" was a territory in need of in-sight/site/cite-ment. Viewed from London, the region was a blank space on a proudly colored British imperial map, a tabula rasa that had remained outside the measurements, classifications, and names that came with European ways of seeing. As an unseen and unmarked region, the territory was an attractive virgin space that had not yet succumbed to the advances of British authority. The purpose of Mackinder's expedition to Mount Kenya was to write on this blank page, to focus a British scientific and imperial eye on the region and thereby in-sight/site/cite it. Mackinder occupied the judicial place of the king in the expedition. He laid down the laws governing the expedition and, together with Hausberg, lorded it over the natives. He was empowered to see, site, and cite the earth by virtue of his race, gender, and status as a representative of the king's own geographical society. His eyes were sovereign, his authority guaranteed by his male body, his white skin, and his European learning.

The expedition was so arranged as to leave Mackinder "free for observation and survey."[19] The results were a diary account narrating the expedition together with an address and appendix, presented to the RGS in January 1900 (about two and a half months after his return), of preliminary notes on the "scientific results" of the Mount Kenya expedition. This appendix comprised a set of cartographic maps, a record of altitude measures (disputed at his address), a sample of rock specimens, a record of glaciers and glaciation, meteorological measurements, a set of photographs, an inventory of zoology and botany (deposited in the British Museum), and observations on anthropology and native place-names. For Mackinder, geographical research was about the measurement and survey of colonial regions, calibration using an imperial scale and observation using an imperial eye. Although only one possible way of writing the earth, Mackinder's geographical eye claimed for itself a totalizing authority and objectivity. Indigenous knowledge of the region had significance not as geography but as curiosity and myth, material Mackinder gathered for an appendix section on "native place-names" (485). Mackinder's science was neutral yet absolutely superior, objective yet thoroughly personalized. By the power of his maps, inventories, photographs, and descriptions, Halford Mackinder took possession of Mount Kenya for British science and the British Empire.

The epistemological assault and capture of Mount Kenya by British geography was an act given meaning by deeply gendered understandings of science and nature. For a start, mounting Mount Kenya was part of the well-established Western tradition of positioning nature, particularly glaciated mountains (sites of purity and beauty — like the Alps — that were the locus of recreational activity for the leisured classes of Europe), as a feminine object to be conquered and dominated. The interior of "British East Africa" is represented by Mackinder as a feminized space to be penetrated, a territory previously breached by other explorers but not yet conquered (453). Mackinder described his expedition as having "a reasonable chance of completing the revelation of its alpine secrets" (454). Commenting on the origins of this understanding of nature, Carolyn Merchant notes how the constraints against penetration associated with the earth-mother image were transformed in the seventeenth century into a mechanistic system of understanding that sanctioned the denudation and stripping of nature.[20] After the scientific revolution, nature-as-woman "no longer complains that her garments of modesty are being torn by the wrongful thrusts of man" but is represented as "coyly removing her own veil and exposing herself to science" (190). Nature had become a mindless and submissive female body.

It was upon this female body of nature that geography came to constitute and know itself as a "manly science."[21] In the nineteenth century, geography was a regime of truth spoken by privileged white European men who conceived of themselves as heroic explorers of a femininized nature. The RGS was an all-male travelers club (founded in 1830) composed of diplomats, colonial administrators, professional travelers, army officers, and naval officials; until 1915 it excluded women from membership. Although Victorian England had a number of female explorers, the election of women as Fellows of the RGS was fiercely opposed by certain members.[22] Mackinder was elected a Fellow of the RGS in 1886 and remained a member for the rest of his life. Climbing Mount Kenya was a rite of passage for him into a position of authority within the RGS and scientific circles within the empire. He had organized the trip because, as he wrote later, "at that time most people would have no use for a geographer who was not an adventurer and explorer."[23] Penetrating the blank virgin land around Mount Kenya, occupying "dark interiors," and conquering the feminine mountain provided him with a certain manly, outdoors credibility among the "lion-hunting" section of the RGS.[24] It enabled him to achieve a certain scientific manhood by allowing him to assume the position of a returning explorer, an experienced authority among men who had ventured forth and disseminated British science on virgin territory.

As the modern discipline of geography reflects upon its history, it is no longer possible to separate that history from questions of gender, race, and empire. Within the last few years the thoroughly imperialist and gendered nature of this history is at last being exposed, unraveled, and deconstructed. An emergent critical literature on imperialist writing outside the discipline of geography and the conceptualization of critical genealogies within the discipline are posing new questions to those who write within the discipline. Such revisionist literatures are helping to open up geographical knowledge and the process of "worlding" to critical scrutiny for the first time. A fundamental challenge for this literature is to document precisely how geography functioned in an imperialist, gendered, and racist manner. Too often there is a tendency to simply pronounce geography historically guilty of these practices without thorough documentation and empirical analysis. Confining the sins of the discipline to the past is a strategy that permits the geographical gaze of the present to avoid critical scrutiny and decolonization.

The purpose of this chapter is to investigate in detail the imperialism of geographical knowledge through an examination of the institutional context and content of Halford Mackinder's writings. It undertakes a deconstruction of how Mackinder privileged the faculty of sight within the newly emergent discipline, and how he elaborated this into a concept of "visualization," which he described as "the very essence of geographical power."[25] The recognition and cultivation of this important faculty of visual imagination was, Mackinder argued, essential to the education of an imperial people. In order to provide a sense of the historical and ideological circumstances within which Mackinder elaborated his theoretical understanding of "visualization," I will briefly review the circumstances which led to the establishment of geography as a university discipline in Britain, and Mackinder's crucial role in articulating the nature of this new discipline. I then address the concept of "visualization" in some detail and document a Rousseauian-style denigration of writing within Mackinder's arguments that made it possible for geographical sight to appear objective and neutral while actually functioning in a Eurocentric manner.

THINKING IMPERIALLY: MACKINDER AND THE ESTABLISHMENT OF GEOGRAPHY IN GREAT BRITAIN

The founding moment of modern Geography was neither academic nor intellectual, but bureaucratic. In 1874 the bureaucracy of the Prussian state decreed that all royal universities should create chairs of geography. This decision, imposed from above upon universities without their input or consent, forced the institutionalization of what was then a popular domain of knowledge that had no essential identity or agreed-upon coherence.[26] The decree led to the establishment of geography as a university discipline first within Prussia, then within all German universities (1883), and finally throughout the rest of German-speaking Central Europe.[27] Geography was considered to be important for general education, so the Prussian state established the discipline within universities to train an army of primary and secondary school teachers. The French state followed the Prussian example. The deployment of newly formalized regimens of learning such as geography within the classroom helped train and discipline students, in states where regional identities were still persistently strong, into official understandings of national territory and sense of place. In the recently unified lands of Germany, this was a particularly pressing concern.

The political and ideological significance of the establishment of geography as a university discipline in Germany and France was not lost on members of Great Britain's Royal Geographical Society. In 1884 the society commissioned one of their Fellows, Scott Keltie, to produce a report on the teaching of geography on the Continent and its implications for the British Empire. The release of Keltie's Report in 1886 became the occasion for a public campaign by the RGS for the promotion of geographical education as a necessary part of public school and university education. Keltie's Report was transformed into a traveling exhibition, while a series of lectures organized by the RGS trumpeted the importance of geographical education in securing commercial and imperial advantage for states like England.[28] All made the case for the establishment of geography as a university discipline within British universities. To become a discipline, however, a collection of practices (geography) had to be made into a distinct and coherent subject with scientific credentials (Geography). This task was a demanding one, for many intellectuals within Britain's universities viewed that which claimed the name "geography" as no more than a branch of geology or history.

In their search for an intellectual booster for their geography-as-Geography campaign, certain figures within the RGS turned to an ambitious and struggling Oxford graduate by the name of Halford Mackinder. Halford Mackinder had come to their attention because of the reported success of his efforts as a lecturer for Oxford University Extension, an adult education program that sought to provide classes for "ordinary people" in the provincial towns of England. Soon after being elected a Fellow of the RGS in 1885 Mackinder was asked to present an address to the society that would outline a "new geography" with a coherent subject matter and purpose. Mackinder's resultant paper, "The Scope and Methods of Geography," was presented to the RGS on January 31, 1887, and was widely discussed within the society and in newspapers such as the *Times*.[29] The concerted campaign by the RGS to push for the establishment of geography as a university discipline led in February 1887 to the announced establishment of a position of reader in geography at Oxford University. In June 1887 Halford Mackinder was appointed to this position. Soon afterward a lectureship in geography was established at Cambridge, again through the intersession of the RGS.[30]

Mackinder's key address to the RGS in 1887 begins with the question "What is Geography?" It was, he acknowledged, a strange one to address to a geographical society. Nevertheless the question was relevant, he argued, for "we are now near the end of the roll of great discoveries." Geography, he suggested, needed to define itself as a "separate sphere of work." Existing geography was riven by a gaping divide between physical and political geography. Knowledge, Mackinder asserted, was one, but "the extreme specialism of the present day seems to hide the fact from a certain class of minds. The more we specialize," Mackinder remarked, "the more room and the more necessity is there for students whose constant aim it shall be to bring out the relations of the special subjects. One of the greatest of all gaps lies between the natural sciences and the study of humanity. It is the duty of the geographer to build one bridge over an abyss which in the opinion of many is upsetting the equilibrium of our culture."[31]

The metaphor of geography as a bridge over various abysses that were said to characterize British culture (and modernity in general) was a popular one within the RGS.[32] It gave the discipline an exceptionalist myth and a consciously political role not simply in academia but in the life of British society as a whole. Mackinder, only twenty-five at the time, concluded his address by stressing the practical value of the discipline:

> I believe that on lines such as I have sketched a geography may be worked out which shall satisfy at once the practical requirements of the statesman and the merchant, the theoretical requirements of the historian and the scientist, and the intellectual requirements of the teacher. Its inherent breadth and many-sidedness should be claimed as its chief merit. At the same time we have to recognize that these qualities will render it "suspect" to an age of specialists. It would be a standing protest against the disintegration of culture with which we are threatened.[33]

Geography offered breadth in a world of insularity and specialization, a bridge in a world of abysses, organic coherence in the face of disintegration and disequilibrium. Implicit within this analysis was Mackinder's evolving organic conservatism, an ideological worldview that led him to read the relative decline of British power in the latter half of the nineteenth century in terms of a disequilibrium and decay introduced into previously harmonious organic communities by unregulated capitalism, technological change, demographic failings, and the general pace of modernity.[34]

The potential of geography to arrest the relative decline of British power and renew the idea of empire was crucial in Mackinder's understanding and elaboration of a new discipline of that name. Addressing the Manchester Geographical Society in 1890, Mackinder chose to speak on a matter of practical interest to the ruling class in the city: commercial geography. Geography, he argued, was not about the collection of useless information about places but a "trained capacity" for thought. This trained capacity was a capacity to picture the world as a dramatic spectacle on a stage. Using the example of wheat from Lahore, the capital of the Punjab, he argued:

If I have been properly trained in geography, the word Punjab will, to borrow a term from logic, probably *connote* to me many things. I shall see Lahore in the northern angle of India. I shall picture it in a great plain, at the foot of a snowy range, in the midst of the rivers of the Indus system. I shall think of the monsoons and the desert, of the water brought from the mountains by the irrigation canals. I shall know the climate, the seedtime, and the harvest. Kurrachee and the Suez Canal will shine out from my mental map. I shall be able to calculate at what time of year the cargoes will be delivered in England. Moreover, the Punjab will be to me the equal in size and population of a great European country, a Spain or an Italy, and I shall appreciate the market which it offers for English exports. This is geographical capacity—the mind which flits easily over the globe, which thinks in terms of the map, which quickly clothes the map in meaning, which correctly and intuitively places the commercial, historical, or political drama on its stage.[35]

At this early stage in his career certain elements in Mackinder's conception of geography were already in place. Geography is about the training of the faculty of sight in a detached pictorialization of the drama of the world. The eye of geographical sight is not an earth-bound eye but an elevated, disembodied, earth-scanning mind's eye that "flits easily over the globe." It is, in short, a panoptic eye. Mackinder's initial RGS address, it must be remembered, concerned the *scope* and methods of geography. It is a panoptically conceived project that seeks to train British schoolchildren, the future guards of the control-tower of the Britain Empire, in a high-altitude way of seeing so the worldwide spaces of empire can be rendered visible, observable, and meaningful to British interests.

Mackinder states that this panoptically trained capacity involves the ability to think in terms of maps and to clothe maps with meaning. The implication that maps are not immanently meaningful themselves but naked texts without meaning is worth noting. It hints at the operation of a system of understandings within Mackinder's text between that which is innocent and transparent (naked) and that which is social and opaque (clothes). Further, the ability to intuitively place entities onto a stage suggests not only that the physical environment is a passive stage upon which an active human drama is played out but that the geographer's position is that of a removed and detached observer of the spectacle of human affairs. That "geographical capacity" might enable one to see unequal exchange or colonial domination is not possible within the scopic regime codified by Mackinder, a scopic regime that configures the world-as-exhibition for European imperial eyes.

Mackinder's most explicit articulation of his understanding of geography's potential worth in renewing the British Empire is found in a 1907 lecture, "On Thinking Imperially." In it Mackinder suggests that perhaps the chief difficulty in the organization of the British Empire is "the difficulty of effectively picturing the idea of that Empire."[36] The observation reveals Mackinder's extraordinary belief in the power of pictorial ideas and his reduction of the problems of empire to questions of seeing properly. It presumes an enlightened community of interest—a "we"—to whom the British Empire is a civilized and desirable form of political organization. The problem, as Mackinder represents it, is that this community of interest is not shared widely enough among the peoples of the British Isles. There are, he suggests, many educated Englishmen who really fail to grasp even the idea of the United Kingdom, but continued to think in terms of England, Scotland, and Ireland and not the United Kingdom as a whole (Mackinder, it could be argued, was also one of them, for he automatically privileged England over all other places). This unfortunate fact, Mackinder notes, is not peculiar to the British Isles, since strong regional identities were also evident in Germany and the United States. Furthermore, one also has the problem of imbuing a single idea to heterogeneous social groups, some well-educated, others half-educated, while some "do not really belong to the present age of the world" (34).[37] The great and "fascinating problem of our time," Mackinder suggests, "is to see whether, utilizing our modern resources of communication—printing, travelling, telegraphy, and the rest of them—we cannot imbue our whole people with the new idea of the British Empire, without waiting for our common defeat and common victory, our Jena and Sedan" (34).

The German reference points mark a preoccupation with German power, a strategic comparison that provokes him to articulate an extraordinarily ambitious socialization project. This project was nothing less than an ideological assault on the minds of British children, a campaign conceptualized as a "liberation" from previous ways of seeing the world. Remnants and residuals of existing territorial identities and subjectivities were to be erased. By laying greater stress on imperial history, "we should seek to emancipate the minds of our children from excessive insularity. By setting them free from what I will describe as the 'Channel prejudice,' we shall enable them to grasp the strategic considerations of a world-wide Empire; and, indeed, the real conditions governing the defence of the home country itself" (36). "It is essential," he adds, "that the ruling citizens of the world-wide Empire should be able to visualize distant geographical conditions.... Our aim must be to make our whole people think Imperially—think that is to say in spaces that are world wide—and to this end our geographical teaching should be addressed" (37–38).

It was to this end that Mackinder worked to establish various institutions and organizations explicitly concerned with the problematic of visualization, geographical education, and empire. In 1893 Mackinder, together with like-minded others, founded the British Geographical Association, an organization orientated toward the concerns of secondary school teachers of geography. The original purpose of the association was to provide for the exchange of lantern slides between teachers and generally to improve the quality of visual instruction in the teaching of history and geography.[38] Mackinder also argued for the establishment of a Geographical Institute in London to match similar research-orientated institutions in Germany. This proposal was modified but eventually resulted in the establishment of a School of Geography at Oxford in 1899 with Mackinder as its head.[39]

In 1902 Mackinder was involved in the establishment of a Visual Instruction Committee of the Colonial Office. This committee was charged with considering what system of teaching to develop with regard to the empire. It began by promoting the use of a small book of lantern slides on the United Kingdom in Ceylon, Malaya, and Hong Kong. Eventually, with sponsorship from the queen, it commissioned a series of illustrated lectures on various parts of the British Empire and sent an artist-photographer (Hugh Fisher, an artist associated with the *Illustrated London News*) to various colonial locations to gather material for illustrated color slides.[40] Mackinder published a collection of lectures on the United Kingdom (1905) and on India (1910) under the auspices of the Visual Instruction Committee, the former in six different editions for various parts of the empire.

Mackinder's greatest efforts, however, in promoting imperialism using geography went into the writing of a series of textbooks on geography for children between the ages of eight and fifteen. Between 1906 and 1914 Mackinder wrote five books in a series called *Elementary Studies in Geography and History* that were used in British classrooms for the following two decades and beyond. The most successful of these volumes, *Our Own Lands* (1906), went through nineteen editions, the last in 1935, while the second, *Lands beyond the Channel* (1908), went through fourteen editions, the last in 1928. Mackinder also published a "practical companion" to these books called *The Teaching of Geography and History: A Study in Method* (1914, with a second edition in 1918) addressed to public school teachers of geography and history. Mackinder's textbooks cover the full range of the secondary educational system, from infant level to the age of fourteen or fifteen. Together they constitute a remarkable attempt to capture the geographical education process within Britain and refine it to serve political purposes.[41]

Mackinder's efforts for geographical education should be understood in the context of larger late Victorian and Edwardian social concerns with "population," "health," and "national efficiency." Foucault has noted the emergence of "population" as an economic and political object of analysis and discourse in the eighteenth century. "Governments," he argues, "perceived that they were not dealing simply with subjects, or even with a 'people,' but with a 'population,' with its specific phenomena and its peculiar variables: birth and death rates, life expectancy, fertility, state of health, frequency of illness, patterns of diet and habitation."[42] By the early twentieth century, these concerns had long shaped state-society relations in authoritarian states like Imperial Germany and Meiji Japan. State-society relations in Great Britain and the United States were different, but in both states strong "health" and "national fitness" movements were reconfiguring how the state conceptualized and treated its inhabitants. Theodore Roosevelt articulated an ideal of "the strenuous life" and manly virtue in the United States. In Britain, the poor health and disastrous performance of British recruits in the Boer War helped galvanize a movement for "national efficiency" among certain sections of the establishment. "National efficiency" was far from being a homogeneous political doctrine or ideology.[43] It brought together a variety of different concerns within the British establishment over questions of poverty, diet, physical degeneration, disease, and the inefficient functioning of the state. Seebohm Rowntree's famous study, *Poverty: A Study of Town Life,* was published in 1901; it linked, for the first time, social policy to the science of nutrition.[44] The effects the slum conditions of British cities had on the physical health of potential military recruits became an object of investigation at this time. A Royal Commission on Physical Training in Scotland held hearings in 1902, while an Interdepartmental Committee on Physical Deterioration sat for six months between December 1903 and June 1904 investigating whether hereditary or environmental factors caused "physical deterioration."[45] "National efficiency," Searle argues, was "an attempt to discredit the habits, beliefs and institutions that put the British at a handicap in their competition with foreigners and to commend instead a social organization that more closely followed the German model."[46] Interpreting "national efficiency" as a rejection of Gladstonian liberalism and a movement toward a more authoritarian model of government underplays, however, the important neo-Lamarckian biological precepts that helped define the very gaze the movement brought to political life. The purpose was not simply reform of the state machinery but a reconceptualization of the nation as an organism that needs to be kept fit and healthy if it is to survive in the competitive environment of the state system. The ambit of the reforming mission of "national efficiency," therefore, extended well beyond the institutions of government into the daily life and habits—nutritional, recreational, sexual, spiritual, and educational—of "the population."

Mackinder's role in championing discourses of "national efficiency" is not a minor one. In England he was among the first to call for the addition of a detailed measure of male demographic health to the traditional calculus of power used by the state. Joining "sea power" and "land power" as objects of measurement, analysis, and speculation was "man power," a term used by Herbert Spenser and taken up by Mackinder in a 1905 address to the Compatriots Club. The deficiencies of a purely quantitative approach to population by the state is the point of departure of Mackinder's address: "You may enumerate a population and set it tersely down at so many millions, but you will not thereby measure the strength of a nation in competition with other nations, for that would involve the assumption that all men are equally efficient." The new orientation proposed by Mackinder involved a turn away from statistical indexes of population size, trade totals, and national wealth to concern with "the output of human energy for which wealth affords but part of the fuel." Mackinder defines human energy not in abstract statistical terms, but in terms of the practical use to which a nation puts its laboring men. His argument is a neo-Lamarckian articulation of Joseph Chamberlain's argument against free-trade political economy and in favor of empire building through imperial preferences. "Let empire-builders show that they value man-power at home and in the colonies more than wealth, and the masses of our countrymen will learn to value the Empire as the protection of their manhood."[47] The best means of doing this is to restrict foreign imports and emphasize manufacturing in Britain and the rest of the empire. National strength, efficient energy use, empirewide manufacturing, and manhood are seamless elements of national strength for Mackinder.

Discourses of "national efficiency" invented a biopower front in Britain's imperialist rivalry with Germany. The outcome of the great struggle between competing world empires (or "world organisms," as Mackinder often termed them) would be determined by which empire made the most efficient use of its reservoirs of man power. It was toward the productive use and training of the future "man power" of Britain that Robert Baden-Powell, a veteran of the colonial frontier and "hero" of the Boer War, established the Boy Scouts in 1908 as an organization designed to remedy Britain's moral, physical, and military weakness.[48] Their motto — "Be Prepared" — revealed an "implicit understanding of the world as a place of ceaseless struggle" in which those who are best prepared and most skilled survive.[49] It is in this context of establishment projects to create trained bodies for empire that we must locate Mackinder's campaign for geographical education. Mackinder's elementary textbooks were being written at the same time as Baden-Powell was organizing the Boy Scouts. *Lands beyond the Channel* was published in the same year as Baden-Powell's manual *Scouting for Boys* (both 1908). Both projects shared certain similarities: a practical guidebook style, a Rousseauian back-to-nature ethos, an emphasis on the study of nature, and a concern to refine techniques of observation and surveillance. Mackinder's efforts to codify a faculty of "visualization" that defined geography, therefore, should be understood as part of a larger regimental offensive directed at the minds of local educators and the imaginations of receptive schoolchildren.

WRITING THE GEOGRAPHICAL GAZE

Mackinder's understanding of education and visualization owe much to the writings of Jean-Jacques Rousseau, and his companion book for teachers using his textbooks explicitly acknowledges Rousseau's *Emile* as the source of "a great many suggestions of a valuable character."[50] Like *Emile, The Teaching* is written as a practical handbook for teachers to assist them in the training of young men.[51] Rousseau's *Emile* outlines a vision of a naturalistic education which places great stress on the importance of a carefully controlled exposure of the young pupil to learning about the world. Failure to adhere to an assumed naturalistic sequence of development can result in the early subversion and corruption of the education of the child by society and its social conventions.

One of these corrupting conventions is the practice of writing, wherein a symbol takes the place of a real thing. According to Rousseau, the young child should not be encouraged to read books: "No book but the world, no teaching but that of fact. The child who reads ceases to think, he only reads. He is acquiring words not knowledge."[52] Rousseau enunciates a general rule to all teachers: "Never substitute the symbol for the thing signified, unless it is impossible to show the thing itself; for the child's attention is so taken up with the symbol that he will forget what it signifies" (133).

The educational principles Mackinder articulates in *The Teaching of Geography and History* are organized on the basis of Rousseau's general rule. No books are to be used in the teaching of geography until the pupil is eight or nine years of age. "The one deadly sin in the early stages is hurry. Just as we may kill individuality by denying the reality of that which is real to the child, so may we kill it at a later stage by substituting mere symbols for real images. In geography and history the besetting symbols are words. Every time that we teach a name with no real image attached we are causing 'little ones to stumble.' "[53] It is certain, Mackinder claims, that "vast numbers have minds that were maimed in childhood by learning the name before the thing" (10). Past English educational methods, he argues, "inverted the natural order" by teaching analysis, abstraction, and reasoning before "the real vivid fact itself" (12). Rousseau's own reflections on the teaching of geography in *Emile* make exactly this point. Geography is part of the "useless lumber of education," a technical discipline that, like languages, teaches knowledge that is remote from the experience of the child:

> In any study whatsoever the symbols are of no value without the idea of the thing symbolized. Yet the education of the child is confined to those symbols, while no one ever succeeds in making him understand the thing signified. You think you are teaching him what the world is like; he is only learning a map; he is taught the names of towns, countries, rivers, which have no existence for him except on the paper before him. (74)

> You wish to teach the child geography and you provide him with globes, spheres and maps. What elaborate preparations! What is the use of all these symbols; why not begin by showing him the real thing so that he may at least know what you are talking about. (131)

The crucial divide for Rousseau and Mackinder is that between the world and books, between that which is present, immediate, and close to the sensory faculties (experience) of the child and that which is abstract, written, nominal, and therefore distant from the mind of the child. Geography, for Rousseau, is textual: it is found in books and consists of a system of maps, names, and symbols, all of which are removed from the immediacy of nature. Mackinder not only accepts this understanding but makes it the very basis of *The Teaching of Geography and History*. Geography, for Mackinder, should be nontextual; it is not about books but about the "real." This puts Mackinder in an awkward position, for he is an author of an extensive set of textbooks and a practical guide to the reading of these textbooks. Like Rousseau, Mackinder is caught in the contradiction of being a tireless professional writer who polemicizes against writing and its artificiality. This contradiction is never explicitly faced by Mackinder, for the distinction between seeing and writing hold together his whole pedagogy. The teaching of geography is not about the teaching of signification (reading and writing) but about the cultivation of visualization. An educated geographer is an observer, not a writer. The essence of "geographical capacity" is the power to visualize, not the power to inscribe. Geography is seeing, not writing.

To comprehend the larger place of this crucial distinction within Mackinder's writing, we must briefly explore the system of under-standings made available to intellectuals like Mackinder in the texts of Jean-Jacques Rousseau. In *Jean-Jacques Rousseau: Transparency and Obstruction,* the Geneva-school literary critic Jean Starobinski describes Rousseau's writings as "a philosophy haunted by the idea that human communication is impossible"(5).[54] Rousseau's many works are united in their persistent search for transparent meaning, for the conditions under which human feelings and human words are united as one. Rousseau anchors his search around an idealized vision of transparency as presence, which is either the self-presence of the "soul" (or "heart") or the presence of nature as an idealized organic community characterized by face-to-face interaction and unmediated communication. From this initial point of origin, Rousseau charts a whole system of attempted and frustrated communication, and de-scribes a world where appearance and reality are rarely the same. At the center of Rousseau's regime of truth is self-present conscious-ness and an idealized vision of humans in nature, a state where pres-ence is full and transparent. Appearance and reality are in perfect equilibrium. Beyond this is the domain of humans in society where false appearances, opaque meaning, and deception are the norm. Truth is measured by its immediacy and proximity to the absolute self-presence of the former. Around this scheme Rousseau constructs a series of histories of origins of human society, civilization, language, and music. As a rule these histories tell the story of the "fall" of hu-mans from the idealized state of nature because of certain flaws, such as pride, in a still perfectible human nature. As Starobinski notes, Rous-seau takes the religious myth of the "fall" and sets it in historical time, which he divides into two ages: "a changeless age of innocence, dur-ing which pristine nature reigns in peace, and an age of historical change, of culpable activity, of negation of nature by man."[55] The story of modern civilization is thus rendered as a tale of progressively increasing opacity, mediation, and alienation of human beings from na-ture, including their own. For Starobinski, Rousseau's social thought was aimed primarily at establishing or reestablishing the sovereignty of the immediate, that is, the primacy of a value upon which dura-tion has no claim. Put somewhat differently, Rousseau sought to reaf-firm presence (immediate, intimate, present) in societies character-ized by pervasive absence (mediation, distance, historical time).

The great significance of Rousseau as a writer, for Starobinski, is that "he invented a new attitude which became that of modern literature": "The writer singles himself out through his work and elicits assent to the truth of his personal experience."[56] In his sketches of a history of logocentrism, Jacques Derrida accords the texts of Rousseau a prominent place between Descartes and Hegel precisely for this reason. Rousseau, for Derrida, articulates "a new model of presence: the subject's self-presence within consciousness or feeling."[57] Rousseau helped bring into existence the space of "inner consciousness" and articulated a mythology of origins and presence that is still influential in contemporary society. "Self-presence, transparent proximity in the face-to-face of countenance and the immediate range of the voice, this determination of social authenticity is therefore classic: Rousseauistic but already the inheritor of Platonism" (138). Taking issue with Lévi-Strauss's anthropology, Derrida traces its origins back to Rousseau: "The ideal profoundly underlying [Lévi-Strauss's] philosophy of writing is ... the image of a community present to itself, without difference, a community of speech where all members are within earshot" (136).

Rousseau's influential picture of an originary, organic community is a socioeconomic expression of the metaphysics of presence, which works by subordinating writing to the presumed authenticity of speech. It is Rousseau's account of the origins of languages and his sustained devaluation of writing (inherited from Plato) that interests Derrida. Writing, we recall, is dismissed in *Emile* as a mere trifle not worth spending time considering. In Rousseau's thought generally, it is seen as a necessary evil, as a dangerous supplement that corrupts the self-present transparency of the state of nature. "When Nature, as self-proximity, comes to be forbidden or interrupted, when speech fails to protect presence, writing becomes necessary. . . . Writing is dangerous from the moment that representation there claims to be presence and the sign of the thing itself" (144). Writing thus has the paradoxical status of being the destruction of presence yet also a necessary supplement in order for presence to attain plenitude. Derrida plays on this ambiguity in the term "supplement" to argue that writing is not a *re-presentation* of consciousness, experience, or speech but a necessary precondition for their existence. The supplement is not additional but fundamental. Rousseau's very notions of self-presence, uncontaminated nature, and proximity require writing, which is understood by Derrida in its broadest sense as all that gives rise to signification (arche-writing). Writing is that which opens up the possibility of consciousness, seeing, knowledge, and symbolic representation. Writing is already there in Rousseau's story of the state of nature and the origins of signification. It is only through its active suppression and denigration that Rousseau can offer his mythical story of origins, self-present consciousness, and transparency.

Derrida's reversal and displacement of Rousseau's understanding of writing has a heretofore unrecognized relevance for Mackinder's understanding of visualization and the geographer's gaze. To document and demonstrate this relevance I have organized the various descriptions of "visualization" throughout Mackinder's texts under three general headings: (1) the apparatus of visualization: the "mind's eye" (I use the term "apparatus" rather than "organ," since the "mind's eye" is understood as a noncorporal observational tower rather than a living and sometimes imperfect human organ); (2) the normative type of visualization; and (3) the place of visualization within the art that is geography. The argument I wish to make is not difficult to grasp. Mackinder's concept of visualization is one predicated on a logocentric occularcentrism. The active suppression of "writing," taken as a mark of the social, is a condition of its existence. Mackinder's texts seek to make the supposed mark of the social disappear and thereby clear the way for claims of access to the natural, the presocial, that which is supposedly unmediated presence. Yet, the very construction of Mackinder's arguments assumes a sociality that is not revealed and acknowledged. Embedded in Mackinder's construction of the natural and presocial is a latent Western system of signification that functions as a universal horizon of intelligibility for all of Mackinder's claims. In other words, the set of understandings that make visualization possible for Mackinder rest on the aporia that seeing is prior to writing qua site/citing. When pressed, however, this asserted priority of seeing over writing begins to disintegrate, and the socially constructed and situated character of Mackinder's construction of sight is exposed. Documenting the hidden sociality of sight enables us to reveal an epistemological imperialism in the functioning of the geographical gaze.

THE APPARATUS OF VISIBILITY: THE MIND'S EYE

"Vision" has long been accorded a special place in Western epistemology. As we have already noted, however, we should not assume
an essential or continuous identity for "the visual" in the Western
tradition. Rather, that tradition has been characterized by a series of
different "scopic regimes," each of which has its ambiguities and
ambivalences. Plato accords sight a particular preeminence in his
writings and speaks of the "eye of the mind."[58] The concept of an
"idea" (*eidos*), for example, is a Platonic term for "things that are
seen." Plato's writings on knowledge and the senses helped delimit
the emergence of a concept of "scientific objectivity" within the
"West." According to Evelyn Fox Keller and Christine Grontkowski,
two features were crucial to this process: (1) the separation of subject from object and (2) the separation of knowledge from the unre-
liability of bodily senses. Plato helped establish the conditions for the
latter more than the former. Vision became more than ocular perception in Plato but a metaphor for disengagement from the body:

> Vision is that sense which places the world at greatest remove; it is
> also that sense which is uniquely capable of functioning outside of
> time. It lends itself to a static conception of "eternal truths." Although
> itself one of the senses, by virtue of its apparent incorporeality, it is
> that sense which most readily promotes the illusion of disengagement
> and objectification. (213)

It is Descartes that helps codify and refine such notions further in
the seventeenth century. The active sense of "seeing" in Plato (his
emission theory) is replaced by an understanding of the eye as a passive recorder of an external world, a type of camera obscura. "Having made the eye purely passive, all intellectual activity is reserved
to the 'I,' which, however, is radically separate from the body which
houses it" (215). An inside "self" contemplates an outside "world."
The "mind's eye" becomes a passive observatory or mirror upon
which this outside world makes its impress. Vision is decorporal-
ized; it is an apparatus, not an organ.

Two points are worth noting about this development. First, the
Cartesian separation of the subject from that which it observes (object) is the condition upon which classificatory projects like "natural
history" are built. Things are separated from words. The observing
rational "I" enables the creation of multiple projects of investigat-

ing, naming, and labeling. Second, in making "reason," "science," and "objectification" disembodied activities, they are gendered as masculine; their outside or other is "desire," "passion," and the "body," positions coded as feminine. The prioritization of the male over the female, the mechanical over the organic, mind over passion, and the objective over the situated are necessary conditions for the construction of any aspiring science (like geography) by the nineteenth century.

It is worth remembering that observation in the seventeenth and eighteenth centuries was hardly a knowledge organized exclusively around visuality. It was only in the early to middle nineteenth century that vision itself became a physiological object of knowledge, part of a larger shift within the empirical sciences toward detailed investigations of the human body and "man in his finitude."[59] The invention, development, and refinement of processes of observation within the empirical sciences (including observations of the eye itself, light, sensory nerves, and vision) normalized a position of observational neutrality from which diagnoses, judgments, and evaluations could be made. For certain empirical sciences and arts in the nineteenth century, the possibility of seeing as a pure activity with its own truth came to define their methodologies. Medical science sought to train its diagnosticians in "pure perception,"[60] while the English art critic John Ruskin defined painting as dependent upon the "innocence of the eye."[61] It was the expressed possibility of attaining a "pure visuality" through professional training that Mackinder (who expressed admiration for Ruskin's ideas) used to define the specificity of geography among the sciences. In a paper to the Imperial Education Conference in 1911 Mackinder represented the matter thus:

> The mind has an eye as well as an ear, and it is possible to train this eye by appropriate methods to as much accuracy and readiness of thought as may be imparted by the ordinary literary methods to the mind's ear. It is, of course, true that many people visualise literature, and *see* the printed page rather than *hear* the voice of the author. Such visualization is, I venture to say, a perversion of literature, and a waste of the visualising power. The music of language was meant to be heard. The power of visualization was meant for real things, rich in shape and color, not for the combinations and permutations of the letters of the alphabet. Let our literary teaching appeal to the

mind's ear, and our geographical and historical teaching to the mind's eye. As Thring of Uppingham had it, "the true geographer thinks in shapes." May I add, "and the true historian in movements"?[62]

The passage contrasts the positiveness of nature to the perversion of the social. Speech is juxtaposed to writing and hearing to reading, with the former as the standard by which the latter should be measured. Speech is held to be equivalent with accuracy and immediacy. The spoken word is a primary signifier; it has a privileged proximity to presence (thought).

Seeing, too, has a privileged proximity to presence, but the innocence of the eye can be perverted by the printed page, the mark of the social. Nature fights the social for the eye's attention. An eye that reads literature is an eye that has lost its natural innocence, its inherent goodness. Mackinder's evocation of Thring of Uppingham in this context is interesting in that one can read the passage as a Victorian morality tale told to justify a regime of "good learning," the very narrative of Thring. The potential perversion of the innocent (the schoolboy in general, the eye in particular) necessitates a regime of disciplined training that orientates the eye toward the natural and the virtuous. Such a regime of training will help safeguard the manly character of the British race. If trained properly, the mind's eye and the mind's ear can both function as natural organs that facilitate the pure movement of presence by means of primary signifiers (spoken words, pictures). In certain other passages Mackinder suggests that the eye is superior to the ear, for its glance instantaneously conveys a picture whereas the ear is confined to the spoken word, which is comprehended only one word or sentence at a time.

Writing, for Mackinder, is dangerous because it is distant and removed from presence. It is a secondary form of signification, the sign of a sign, the signifier of the primary signifiers. That the letters of the alphabet are considered perverting and the opposite of "real things, rich in shape and color" is indicative of writing's distance for Mackinder from presence and authenticity.

Coexisting with such a set of categorical utterances in the texts of Mackinder, however, are a subordinate set of tentative utterances that undermine and compromise the rhetorical force of that which is categorically argued. The Rousseauian separation of "real things" from words is persistently problematic, since the recognition of "real things" with the attributes of "shape" and "color" is dependent upon a language that has a concept of "the real" and "a thing," together with a shared social understanding of how one recognizes "shape" and "color." In more provisional utterances Mackinder's texts allow a certain slippage in his claims. In the chapter "Thinking in Shapes" in *The Teaching of Geography and History,* Mackinder acknowledges that thought cannot advance except by the use of words. This offhand remark radically undermines Mackinder's ostensible attempt to keep experience, thought, and the operation of the senses (seeing/sighting) separate and distinct from formal learning, words, and social conventions (writing/citing). If advanced thought cannot be separated from words, then "thinking in shapes" cannot proceed without reliance on language. If thinking is reliant on language, then it is compromised, for it inevitably becomes mired in the corrupting social conventions of representation.

The ostensible figuration of Mackinder's texts, however, works to keep such readings at bay. The above acknowledgment is immediately followed by the observation that "it is none the less true that the earliest stages of thought are probably without words, and it is certainly true that when words have done their work the most rapid and comprehensive stretches of thought are with the mind's eye rather than with its ear."[63]

The geographer, however, inevitably writes. Mackinder, like Rousseau, treats the task of writing as a supplementary activity to that which is primary (observation, visualization). When writing, the geographer should avoid terminology and attempt to compose in a neutral and plain manner. Words are tools that should be used in a mathematically precise and workmanlike fashion by the geographer.[64] Mackinder writes that he would have the young geographer "practiced in the use of an almost Ruskinian, purely descriptive language, with terms drawn from the quarryman, the stonemason, the farmer, the alpine climber, and the water-engineer."[65]

The possibility of a "purely descriptive language" is crucial to preserving the neutrality and objectivity of geography. Interestingly, he associates purely descriptive language with outdoor activities and livelihoods, the language of those whose everyday experience is close to the rocks and fluids of nature.

PRIMITIVE VISUALIZATION: THE CHILD AND THE SAVAGE

The type of visualization to which the geographer should aspire is primitivist and natural. It is the type found among those closest to an original state of nature: children and savages:

> I plead for the cultivation in geography and history of that visualising power which in rudiment is natural to the child and the savage, but which tends to wither rather than expand in the presence of the printed page and of the ribbons of landscape seen through the windows of a railway carriage.[66]

Children and savages were of a kind because of their assumed externality to advanced society and social conventions. Both were united by their proximity to nature: childhood was the age of immediacy while primitive peoples lived in a condition of immediacy and transparency. Both child and savage are considered to be without the corruption of writing. Opposed to these state-of-nature subjects is the decay and pace of modernity, represented here by writing, which induces withering of the visualizing powers, and the speed of the railway. Both writing and the railway signify distance from the presence of nature.[67]

This rudimentary and uncontaminated natural visualization must, as we have already noted, be cultivated in a proper manner. This is the task undertaken in *The Teaching of Geography and History*, which begins with a paean to the natural faculties of the child. Among these is an active roaming imagination with a strong propensity to play games of "make-believe." The child's supposed ability to move freely between the inner and outer world, or, expressed negatively, its inability to distinguish daydreams from facts and the imagined from the real, is considered a cardinal virtue by Mackinder. Indeed, Mackinder extends this set of claims to suggest that intellectual power in adult life comes from childlike qualities:

> Almost all fine origination in the adult comes from the power of thinking in images, or as we say, of visualization. Constructive genius lies in the child-like power of seeing "what is not" continued without break from "what is," or in other words, of piercing through the material into the immaterial world. The prophet, poet, philosopher, and scientific discoverer have "insight" for their essential power, or the power of reading more into facts than is obvious to the plain man.[68]

Positive visualization, therefore, is the power of thinking in images. Images and pictures are equivalent. Rousseau writes in *Emile* that before the age of reason the child receives images, not ideas. Images are "merely pictures of external objects, while ideas are notions about those objects determined by their relations." Images can exist by themselves in the mind, whereas ideas are always connected to other ideas. "When we image we merely perceive, when we reason we compare. Our sensations are merely passive, our notions or ideas spring from an active principle which judges" (71–72). Mackinder described the ability to think in pictures as "the essence of all fruitful thought."[69]

Besides, and related to, the ability to "think in images" is the ability to "make believe," which is described by Mackinder as a type of travel beyond. This can be either travel beyond tactile, substantive domains into nonempirical domains or, as Mackinder describes elsewhere, travel beyond the horizon of human vision. Mackinder likens the training of the child's mind to flight: "We have to take him on the wings of his facile imagination...into the regions he cannot even see" (3). Both geography and history, he adds, involve "the picturing of absent things and movements, some absent in space and some in time, but most in both." Through imaginative flight absence can supposedly be made present to the mind's eye. It is the business of the teacher to "lift our pupils to higher points of view, so that in geography they may see beyond the horizon." Those who stay on ground think parochially and "have never learned to thrust and joy in the power of their mental wings" (111). The geographer's gaze thus extends beyond the local to a visualization of things that are not strictly visible and immediately present to the eye. The trained geographer sees things that are invisible to the ordinary citizen, for his faculties leave their earthly abode and begin to imaginatively "flit" across the globe.

Mackinder's celebration of an imagined naturalistic seeing leads him to positions that are more than paradoxical, given his political commitments and professional practice (especially in British East Africa). First, the purpose of fostering the geographical faculty of visualization in the British educational system is to produce a British race fit to govern as imperial masters. Yet the very practice of visualization requires that this same British race begin to see and think like savages! The salvation of the British Empire required a savage sensibility. This remarkable irony points to larger tension in the historical forms of Western colonial discourses. The myth of the "noble savage" is a product of the very modernity Mackinder seeks distance from, a constructed signifier that became a locus for expression of the ambiguity of the Enlightenment. The "savage" is the outside, the other by which European colonial discourse consolidated its own enlightened subject position as a universal standard and naturalized its own authority to rule over colonized territorities (like British East Africa). The ennobling of the "savage" is merely a different variant of the operation of this hegemonic system, a move that distances Europe from its own savage conquests in the colonies yet reinscribes the universal European subject and provides renewed legitimation for continued European rule. Pratt uses the term "anti-conquest" to refer to the strategies of representation whereby European bourgeois subjects sought to secure their innocence in the same moment as they assert European hegemony.[70] The writing of the noble savage in popular Romantic literature is undoubtedly part of this strategic movement. In inventing the naturalistic space of the "noble savage," Europe was cleansing itself of its own "heart of darkness" while also reinventing its imperialist project. The savage was noble yet also dangerous. Furthermore this figure was represented even by the best anthropologists as being without writing, a belief, as Derrida argues, that rests on a peculiarly Eurocentric conception of writing.[71]

Second, Mackinder's professional commitment to a detailed program of pedagogy operates in an uneasy tension with his ostensible arguments about primitivism. Mackinder, like Rousseau, offers an adult's version of childhood as an idealized, uncontaminated state of grace before the "fall" that is socialization. But what Mackinder proposes is a socialized primitivism, an intellectualism built on an anti-intellectualism. In practice, his argument is a romanticization of disciplining, a program of regulated training that is all the more pernicious for its thinking of itself as naturalistic. The mind's eye is not left alone but explicitly trained and socialized into a particular way of seeing (sighting sites/placing places). The geographical observer is, in practice, made into an observer of rules, one who observes a particular system of epistemology and the possibilities it allows. To learn geographical observation, therefore, is to learn a way of writing the world.

These two tensions in Mackinder's arguments are more than paradoxical. What they disclose is the operation of a critical blindness in Mackinder's texts, an aporia that makes his narratives possible. Primitive innocent seeing is supposedly a capacity that is prior to writing. Yet, whenever Mackinder describes this naturalistic form of seeing, he is forced to use understandings that require a system of signification already in place. To hold to the possibility of "thinking in images" or "thinking visually" presupposes a system of codes that allows one to designate certain objects as "images," "visual pictures," or "words." To hold to an epistemological stance structured around the innocent perception of external objects requires that one already operate with an epistemology that differentiates between the self and the world, consciousness and objects, an inside and an outside, passive and active, subject and object, and so on. The very idea of innocent perception, therefore, is already a corrupted interpretation. Mackinder's texts are blind to that which makes sight possible, to the codes of signification that designate a field of vision and establish conditions of visibility, and to the rules of administration governing objects, events, and processes within this field. Seeing, in sum, is reliant on writing qua signification. We can document the operation of this strategic aporia by considering Mackinder's descriptions of the techniques that constitute geography.

VISUALIZATION AND TECHNIQUE: DRAWING, MAPPING, AND READING NATURE AS A SILENT MUSICAL SCORE

Throughout his writings Mackinder frequently stated that geography is a science, an art, and a philosophy.[72] Geography is a science because it is concerned with observation and measurement; it is an art because the root of the discipline lies in "freehand" drawing and skill with a pencil;[73] and it is a philosophy because it provides the training necessary to supply values for sound practical judgments. How Mackinder understands its status as an art is worth exploring. In 1895 he offered the following "sketch" (his term) of an ideal geographer. It is long but worth quoting in full, for it brings together so many different themes within Mackinder's texts:

He is a man of trained imagination, more especially with the power of visualizing forms and movements in space of three dimensions — a power difficult of attainment, if we are to judge by the frequent use of telluria[74] and models. He has an artistic appreciation of land forms, obtained, most probably, by pencil study in the field; he is able to depict such forms on the map, and to read them when depicted by others, as a musician can hear music when his eyes read a silent score; he can visualize the play and the conflict of the fluids over and around solid forms; he can analyze an environment, the local resultant of world-wide systems; he can picture the movements of communities driven by their past history, stopped and diverted by solid forms, conditioned in a thousand ways by the fluid circulations, acting and reacting on the communities around; he can visualize the movement of ideas and words as they are carried along lines of least resistance. In his cartographic art he possesses an instrument of thought of no mean power. It may or may not be that we can think without words, but certain it is that maps can save the mind an infinitude of words. A map may convey at a glance a whole series of generalizations, and the comparison of two or more maps of the same region, showing severally rainfall, soil, relief, density of population, and other such data, will not only bring out causal relations, but also reveal errors of record; for maps may be both suggestive and critical. With his visualizing imagination and facile hand, our ideal geographer is well equipped, whether he devote himself to a branch of geography or to other fields of energy. As a cartographer he would produce scholarly and graphic maps; as a teacher he would make maps speak; as an historian or biologist he would insist on the independent study of environment instead of accepting the mere *obiter dicta* of the introductory chapters of histories and text-books; and as a merchant, soldier, or politician he would exhibit trained grasp and initiative when dealing with the practical space-problems on the earth's surface. There are many Englishmen who possess naturally these or compensating powers, but England would be richer if more such men, and others besides, had a real geographical training.[75]

The passage is perhaps the most coherent summary of Mackinder's understanding of geography, the skills that characterize it, and their relationship to the practical tasks of education, economic development, and colonial administration. What is of immediate interest here is the series of connections made between seeing, drawing, and mapping. Disciplined perception leads to the construction of a picture (sight/site) of nature in the mind's eye. The geographer attempts to recreate this picture by (1) producing drawings in the manner of a landscape artist and (2) constructing maps that are nonlinguistic pictures. Nature is represented as a silent musical score that is of a kind with maps. The ideal geographer "finds the same joy in the contoured map *without names* that a born and trained musician finds in the silent reading of a musical score."[76] "Just as the musician can silently read his score and hear the music, so can the geographer read his map and see the picture."[77]

There are many different elements of logocentrism at play here. The attempt to distance the mark of nature from names is an attempt to suggest that nature is a natural text that precedes the written texts found in society. The silent text of nature does not need to be decoded since it is pure presence. The way in which the artist and geographer perceive or the way in which the musician and geographer read is also represented as being external to signification. The mind's eye of the artist, geographer, and musician, the practice of freehand drawing, the construction of a map, the reading of a musical score: all are represented as natural activities uncontaminated by the need to use names. All have a proximity to presence and an immediacy that is disturbed by the use of words, sentences, and narrative. The painting, the map, and the musical score are equivalent nonsymbolically coded texts.

The logocentrism of this system of understanding, however, is an unstable one that is subverted from within by its own reliance on metaphors of inscription (drawing, mapping, and reading). These reintroduce the problematic of signification and inscription into what is supposed to be natural, presymbolic communication. First, it is assumed that freehand drawing is not a social form of writing. Yet freehand drawing works in line, plane, and volume, all socially coded elements of inscription. Furthermore, the activity of drawing pictures is one that is profusely regulated by certain conventions and assumptions concerning perspective.

Second, like artistic inscriptional practices generally, the production of "maps" within the West is a predicated on a whole series of rules, conventions, and symbols that work to define that which is taken to be a "map." David Wood identifies ten separate codes that work to constitute Western maps. Five "extrasignification" codes govern how maps are exploited in society (thematic, topical, historical, rhetorical, and utilitarian), while "intrasignification" codes

(iconic, linguistic, techonic, temporal, and presentational) help structure the internal environment of the map.[78] A map, therefore, is never
without names but is a system of signification, a genre of writing,
an already encoded surface.[79] Furthermore, the limits of a "map"
can never be completely fixed. Any act of signification is a potential
map. To write is to draw a map.[80]

Third, the metaphor of nature as a music score subverts the very
meaning attributed to it by Mackinder. The image of the musician
hearing music when his eyes read a silent score is meant to suggest
the immediacy of the supposed nonverbal communication between
the geographer's eye and the natural text of nature. Yet the uncontaminated purity of this communing is dependent on the reading of
a mathematical code iterated in a set scale with a predetermined
tempo (a musical score). Nature is approached through a highly regulated, artificial language of musical codes. What is presented as unmediated turns out to be highly structured and codified. Indeed, the
symbolic language of musical notation that makes up a musical score
is a particularly Eurocentric text. Produced in the most developed
of societies, it is far from being a natural language. Furthermore, a
musical score is an open-ended text that is subject to more than one
performative interpretation. It can never be held to yield forth an
unchanging symphony of sounds.

In all three cases, signs in the narrow guise of words are condemned and made to disappear, only to be smuggled back in again
as pictorial, cartographic, or musical signs. The effort to ensure that
representation is pure, natural, and innocent is subverted by a necessary reliance on the social codes of signification that operate within
the languages of art, cartography, and music. Mackinder's natural
languages turn out to be highly social. His attempt to represent the
visualization of the geographer as a form of natural and unmediated
communing with nature, therefore, is dependent on a sustained suppression of writing throughout his work. Logocentrism is the condition of possibility of geographical visualization. It is that which
enables it to work imperialistically.

IMPERIAL INCITEMENT

To contemplate the imperialism of Mackinder's account of geographical sight is to contemplate the imperialism of Western forms
of see(d)ing territory, where the eye is a pen and also a penis substi-

tute (the eye that can penetrate territorial interiors).[81] Imperialism, as we have already noted, is act of geographical violence. It begins with the erection of epistemological systems that represent those spaces beyond the familiar and domesticated as blank and virginal. It proceeds to take possession of these spaces and fill them with its own seminal knowledge. Finally, it justifies its possession of these spaces in the name of its own systems of author-ity: the King, Empire, Race, Man, and Science. By in-seminating and in-sight/site/cit-ing a "world" of territorities made in its own image, European science comes to recognize and realize itself.

But imperialism is more than a gendered epistemological phenomenon. It is an economic, social, and cultural phenomenon, a practice that operates throughout a social system and not simply in those spaces designated as colonial. Halford Mackinder sought scientific legitimacy in an act of imperial in-sight/site/citement in "British East Africa," but the territory he spent his professional life trying to colonize was the mind of the British public. Self-conscious, as we have seen in chapter 1, of living in an epoch of closed space, Mackinder understood that the imperial project was now the struggle for relative national efficiency among the Great Powers. He, therefore, became a firm supporter of Chamberlain's incitement to a new form of economic and social imperialism based on imperial preferences and social reform to improve the lot of the British working classes.

Geography, for Mackinder, was, above all, an instrument of social imperialism, a domestic force in a renewed incitement to empire. Mackinder's geography was, first, an incitement to a new imperial identity. He wished to use geography to inculcate an imperial subjectivity, to make the future cadres of empire think of the British imperium as the white man's inheritance and collective responsibility. Mackinder's geography was, second, an incitement of an imperial imagination, a challenge to ordinary British people to think of their interests in global terms. Mackinder's social imperialism led him to speculations that would later became known as geopolitics, a practice of visualizing and surveying global space with a view that writes. He yearned for this to be a popular mode of thought within British ruling circles. Finally, Mackinder's geography was an incitement to an imperial biopolitics, to the surveillance, administration, and proper health of the working classes, so the nation's man power would be fit to fight for the empire. Not only were the minds of the masses to be colonized but so also were their daily habits and bodies.

In August 1914 the first generation of those who were the object of this imperial incitement were called upon, by Mackinder and the rest of the British establishment, to enlist to defend Great Britain's interests. Mackinder devoted himself to the cause of recruiting volunteers, particularly in Scotland, where he spoke at public rallies.[82] Millions responded to the incitement to war from all parts of the empire. On the fields of Flanders and in the hills and valleys of the Somme, these incited young men experienced a very different imperial geography from that taught to them in classroom textbooks. The myths of empire met the mud of battlefields. A bloody geography of trenches, tunnels, "no-man's-land," and "the front" was etched into the earth. Imperial vision was dethroned in this bewildering landscape of blood, sky, and death. Indistinguishable shadowy shapes, camouflaged machines of war, blinding flashes of light and fire, and clouds of deadly poisonous gas produced a disorientating phantasmagoric fog of war that would indelibly imprint itself upon the scopic imaginations of the new century.[83] Imperial incitement led millions of young men to their deaths in 1914; it made the blind slaughter that was the First World War possible.

4

"It's Smart to Be Geopolitical"
Narrating German Geopolitics in U.S. Political Discourse, 1939–1943

Geopolitics: The lurid career of a scientific system which a Briton invented, the Germans used and Americans need to study
— TITLE OF *LIFE*'S MAIN ARTICLE, DECEMBER 21, 1942[1]

In late 1939 a new word forced its way into the ordinary language of American political culture. That new word was "geopolitics," a term already familiar to certain geographers and now propelled to prominence by the outbreak of war on the European Continent. The German invasion of Poland and the subsequent conquests by the Nazi war machine on the Continent created, in the popular mind, an overwhelming need to explain the motivations and design behind the conquests of the Third Reich. Soon after the shock of the Nazi-Soviet pact and the partition of Poland and the Baltic republics, geopolitics made its appearance as the "hidden logic" behind Nazi foreign policy. Geopolitics was a new German superscience associated with the mysterious figure of "Major General Professor Doktor" Karl Haushofer,[2] who was rumored to be the scientific brain behind Hitler's policy of lebensraum.

Even before the outbreak of war, popular journals and newspapers in the United States exhibited a sensationalist interest in the Third Reich. Stories of Nazi spies and fifth-column forces operating in the Americas reflected a general fear yet also fascination with the Nazis. After Western Europe fell to the Nazi *Blitzkrieg*, Americans

111

became even more convinced that the Nazis were using secret plans and internal spy units to achieve their military aims. Hollywood began to make movies about Nazi spies in America, such as the Warner Brothers feature *Confessions of a Nazi Spy* (1939). Many newspapers ran articles detailing the working of German propaganda operations and fifth-column forces in the United States. The climate of panic and fear became so pervasive that J. Edgar Hoover, who in 1940 preached extreme vigilance against fifth-column forces, reversed himself and warned against "cooked-up hysteria" and "fearmongers" who, in some instances, had formed vigilante groups to hunt down German sympathizers and Nazi spies.[3]

The American press's discovery of and speculation on "geopolitics" occurred within the context of these widespread cultural fears and fantasies about Nazi Germany. In 1942 a plethora of books and articles on geopolitics appeared, firmly establishing the term in U.S. political discourse. Commenting on the fad, *Life* magazine described geopolitics as a "five dollar term," a word with a "sinister glamour" that suggested "dark plots, evil intrigue and black magic."[4]

If geopolitics drew some of its appeal from suggestions of its uncanny predictive powers, this did not render it a dangerous irrationality. While German geopolitics was sometimes described in those terms, the conventional opinion was that geopolitics was something Americans needed to know. Geopolitics was a new form of global thinking, an intellectual doctrine that the citizens and strategists of any aspirant Great Power needed to take seriously. Summarizing this attitude in a straightforward and pithy manner, Robert Strausz-Hupé, an émigré from Vienna, declared in a review of works on geopolitics in 1943, "It's smart to be geopolitical." Commenting on the boom in literature on geopolitics, he noted that "the awakening of the American public to global consciousness created a ready market for 'systems' of global politics," and, "in the absence of any similar competitive product, geopolitics became the raging fashion."[5] Strausz-Hupé's analysis of this raging fashion in political discourse was hardly innocent, for he himself was deeply involved in selling this "competitive product" to the American masses.[6] In 1942 Strausz-Hupé published a "minor best seller," *Geopolitics: The Struggle for Space and Power,* with the implicit purpose of forcing a reluctant United States into World War II.[7] After the war, he would push a smart new anti-Soviet American geopolitics and, once the Cold War began, institutionalize its distribution and sale with the creation of the Foreign Policy Research Institute and the policy journal *Orbis*. For Strausz-Hupé and the caste of specific geopolitical intellectuals like him whose careers were made by World War II and the subsequent Cold War, it was indeed smart to be geopolitical.

This chapter is an exploration of the surface of emergence, authorities of delimitation, and grid of specifications of "geopolitics" as an intellectual concept-product in U.S. political discourse during World War II. These concepts are taken from Foucault's attempt to account for the formation of objects in a discourse.[8] Surfaces of emergence, for Foucault, are the immediate domains within which an object emerges. In the case of geopolitics, these domains range from war colleges to situation rooms to newspaper editorial rooms. The authorities of delimitation are the group of professionals recognized as authorities on the object by public opinion and the state. They delimit, designate, and name geopolitics as a particular type of object and conceptual practice. In the case of geopolitics, most of its commentators and champions were Central European émigrés like Hans Weigert, Frederick Schuman, Robert Strausz-Hupé, and Andreas Dorpalen. The grids of specification are the systems according to which the different kinds of geopolitical objects (lebensraum, Monroe Doctrine, land power, and so on) are divided, contrasted, related, regrouped, and classified as elements in geopolitical discourse (41–42). It must be remembered that Foucault's attempt to produce an archaeology of knowledge was later recast by him in genealogical terms.

This chapter anticipates a more comprehensive genealogy of the transformation of "geopolitics" from a specialized and relatively unqualified object at the beginning of the war to a household name for a hardheaded, realist Cold War foreign policy in the 1950s. Restricting myself to the years 1939 to 1943, I consider only the main civil society accounts of German geopolitics in the United States during these years, mostly printed but some cinematic. The archives of the U.S. government (the State Department and intelligence services), which contain accounts of German *Geopolitik* and its relationship to Nazi foreign policy, are left out of this study.[9] In order to trace the range of representations of geopolitics in U.S. civil society during the war, I first consulted two bibliographies of geopolitics to select relevant articles and books.[10] I then followed up on the references in these and other works to establish the literature that would be the basis of this chapter.

To provide an incipient wartime genealogy of geopolitics, I wish to address the discursive construction of a regime of truth about German geopolitics. My approach, therefore, goes beyond the recent revisionist literature on German geopolitics, which is concerned with disclosing the "real" history of German *Geopolitik* and fantastic distortions of it.[11] However, a good part of the "truth" of geopolitics, in a genealogical sense, is its fantastic narration. In fact, it is precisely the exaggerated and illogical quality of many accounts of German geopolitics in U.S. political discourse that interests me. Why, for example, were hysterical and paranoid readings of Haushofer as the man behind Hitler produced in U.S. political culture? What do accounts of an imaginary Geopolitical Institute in Munich run by Haushofer reveal about unconscious fears and fantasies associated with the specification of "geopolitics" as an object in U.S. political discourse? My interest, in short, is in the politics of the construction of a mythology or imaginative economy around geopolitics-as-object (a concept, but also a focus of desire) rather than in revisionist demythologization.

In considering the various ways in which "geopolitics" was narrated in U.S. political discourse between 1939 and 1943, I have found it useful to distinguish between a popular and a middle-brow genre of narrating geopolitics. I use the popular description to refer to short, largely propagandistic pieces on geopolitics found in mass-circulation newspapers, news magazines, and journals like *Current History, Life, New Republic,* and *Reader's Digest,* as well as film. The middle-brow description refers to the middle-brow book market. It encompasses those works addressed to the mass-market-book-buying general public, the public intelligentsia. Books published by nonuniversity presses that get reviewed in *Time* or *Newsweek,* therefore, are part of this heuristic category. More specialist academic works on *Geopolitik* were also published during the later stages of World War II in the United States, but these are not considered here.[12]

The first section provides a brief summary of each genre and some of the key pieces within each. The second section then considers the call for an American geopolitics and why it was considered savvy to be geopolitical. The final section reflects on the dialectic of enlightenment at work as geopolitics emerged as a concept that would be crucial to Cold War strategic and popular culture.

STORIES OF A NEW GERMAN SCIENCE

The Popular Narrative

As might be expected, the Nazi invasion of Poland in August 1939 and the subsequent declarations of war generated considerable speculation in U.S. political discourse in the fall of 1939. What did the war mean? How would it end? What were the ultimate goals of Nazi foreign policy? It was speculation on the latter question that led to the introduction of Karl Haushofer and his son Albrecht to the American public in November 1939 in Henry Luce's *Life* magazine. In an unattributed article illustrated with photographs, Karl Haushofer was introduced to the American public as the "philosopher of Nazism," "Germany's brain truster," the "inexhaustible Ideas Man for Hitler, Hess, von Ribbentrop and the inner elite of the Nazi Party."[13] The photographs evoke a system of signs that soon became standard elements of the iconography of geopolitics: a contemplative elder, Karl Haushofer, on a park bench; a younger Albrecht Haushofer with atlas in hand gazing at a globe while a picture of Rudolf Hess hangs on the wall; an imposing but partly obscured photograph of an impressive neoclassical building, labeled the "great German Academy in Munich"; a global map showing where Haushofer sends his German Academy books; a family shot of the Haushofers from World War I showing the dominating presence of Karl in full military uniform (see Figure 6); a picture of Haushofer's desk with a bust of Napoleon and two detail shots, one of an inkwell in the form of a globe (from Albrecht's desk) and another of small photos of Albrecht with Hitler and Hess in profile (Karl's desk). The gray eminence, buildings where secretive plots are hatched, the globe gazing/conquering fetish, the militarist tradition, and personal influence are the key archetypes in the German geopolitics story.

The article claims that Nazis sit at the feet of Haushofer in Munich: "There is more tall talk in Karl Haushofer's German Academy in Munich than in any other place in the world. The world is remade every day between breakfast and dinner. And this great mass of bombast, couched in German polysyllables, is piped directly to the highest councils of the Nazi leadership." Following this is an account

Figure 6. Patriarchal order. The Haushofer family photographed during World War I, including sons Albrecht and Heinz and Haushofer's "Jewish wife," Martha Mayer. Note Haushofer's knife. Reproduced by permission from Life, *November 20, 1939. Copyright Time Inc.*

of Haushofer's career that is blatantly untrue (it states, for example, that he invented the word "lebensraum") and highly exaggerated (he is the author of more than fifteen hundred books and pamphlets!). Haushofer, it concludes, "has an absolute hold on the Nazi leadership."

The *Life* article on the Haushofers is exemplary of the popular narration of geopolitics in U.S. political discourse that developed from 1939 onward. This genre of narration has three distinguishing elements: first, geopolitics is specified as the key to understanding Nazi foreign policy aims; second, the figure of Karl Haushofer, acting in the role of a behind-the-scenes superbrain or scientist, is key to the genre; and third, the analysis of Nazi foreign policy mixes fact and fiction to create a sensationalist, hyperreal narrative around Hitler and Nazism.

A second celebrated example of such a writing of geopolitics is the article "The Thousand Scientists behind Hitler" in the *Reader's Digest* of June 1941.[14] Frederic Sondern Jr., who also wrote for *Life*, composed this article, which is perhaps the best-known source of the myth of a Geopolitical Institute in Munich led by Haushofer, with more than a thousand scientists, technicians, and spies who formulate the ideas, charts, statistics, and plans that dictate Hitler's every move:[15]

> Here is an organization for conquest, a machine for scientific planning, which no conqueror before Hitler ever had at his command. Ribbentrop's diplomatic corps, Himmler's Gestapo, Goebbels' propaganda ministry, Brauchitsch's army and the Party itself are but instruments of this superbrain of Nazism.

Doctor Haushofer and his men, according to Sondern, dominated Hitler's thinking. Haushofer taught the hysterical, planless agitator in a Munich jail to think in terms of continents and empires. He "virtually dictated the famous Chapter XIV of *Mein Kampf* which outlined the foreign policy Hitler has followed since to the letter." Once in power, Hitler funneled unlimited funds to Haushofer, who established a Geopolitical Institute of experts to produce a vast strategic index of the world:

> The Strategic Index tabulates every phase of every nation's life — every detail of its military, economic, and psychological strength. An impending famine in China, the formation of a new political party

in Argentina, the religious sensibilities of the Panamanians, the personality and tastes of Colonel Batista, the morals and corruptibility of a customs inspector at the port of New York, the views of an important American labor leader—all these are important factors in General Haushofer's calculations.

Although the empirical validity of Sondern's claims are bogus, the narrative register of his account is nevertheless significant. The story of Nazi geopolitics is the story of a superefficient modernity gone wrong, of manipulative omnipotent scientists, an omniscient espionage system, and extreme instrumental reasoning. Nothing happens by chance in this nightmarish modernity; every piece of information is used, every move is part of a grand strategy, a hidden blueprint. Sondern contrasts the resources of the Geopolitical Institute to the limited personnel and funds of the U.S. "observation system," which, while good, is not comparable to that of the Germans.

Sondern's Geopolitical Institute was evoked the following month in an article by Albert Horlins in the *New Republic*. He declared that Haushofer's ideas were marching with the Panzer divisions in Russia. The fact that Operation Barbarossa contradicted the logic of Haushofer's whole position was not noted.[16] Once again, the image of an omniscient brain and institute was foregrounded. Haushofer "carefully charts every cultural, language, racial or other minority, every German, Spanish or Portuguese descendant, and every military and economic weakness in the world, including South America and the United States." "Haushofer's Geopolitical Institute at Munich is a propaganda bureau as well as the core of the best military intelligence service in the world." "The geopolitical planners, like the Nazi carpenters, are building for a thousand-year rule."

Such glib readings of Nazi foreign policy were echoed by certain elements in the U.S. military. Colonel Herman Beukema, a West Point lecturer and leading advocate of geopolitical training for the U.S. military, declared that history would rate Karl Haushofer, prophet of German geopolitics, as more important than Adolf Hitler, because Haushofer's studies made possible Hitler's victories both in power politics and in war.[17] Leonard Engel, writing in the *Infantry Journal*, described Haushofer's "lavishly endowed institute as one of the most efficient information-gathering machines ever devised."[18] Berlin's policy to date had, he argued, followed faithfully in the direction laid down in Haushofer's books and in the *Zeitschrift für Geopolitik*.[19]

This conspiratorial representation of German geopolitics was given cinematic form in 1942 by the U.S. Army propaganda film unit established by Army Chief of Staff George C. Marshall under the supervision of the successful Hollywood film director Frank Capra. The use of film for geopolitical purposes was a new vector of geopower that the Nazi state had already harnessed to project idealized images of Hitler and Aryan identity, most famously in Leni Riefenstahl's *Triumph of the Will* (1935).[20] The unit Capra headed, the 834th Photo Signal Detachment, set about producing a seven-part series of orientation films, *Why We Fight,* designed to educate and motivate Allied soldiers about the war. The first in the series, *Prologue to War* (1942), framed the entire war as a struggle between a free and a slave world, an argument given powerful visual form by graphic sequences showing a world divided into two hemispheres, a hemisphere of dictatorial darkness and a hemisphere of freedom and light. Floods of black ink pour over territories penetrated by thrusting Nazi arrows.

The second in this series, *The Nazi Strike* (1942), explains Hitler's goals by means of a short segment on geopolitics. Accompanying film sequences of busy researchers sorting through index cards, carrying files, and examining a photograph with a magnifying glass, the narrative proclaims: "Set up at Munich was an institute devoted to the little-known science of geopolitics, vaguely defined as the military control of space. Germany's leading geopolitician, a former general, Karl Haushofer, was head man. Here was gathered together more information about your home town than you yourself know." Representing itself as exposing what was in Hitler's mind as he stood before a mass rally in Nuremburg, the film explains that geopoliticians see the world's land as divided into a one-third Western Hemisphere and a two-thirds World Island, which holds tremendous resource wealth and seven-eighths of the world's man power. During cartographic sequences of a swastika marked Germany flooding out across first a European map, then a Eurasian map, and finally a world map, the narrator declares: "Hitler's step-by-step plan for world conquest can be summarized this way. Conquer Eastern Europe and you dominate the Heartland. Conquer the Heartland and you dominate the World Island. Conquer the World Island and you dominate the World."[21] As the world map is overlaid with a huge swastika, this is proclaimed to be Hitler's dream at Nuremburg. Although

Prelude to War was offered free of charge to movie theaters in 1943, it did not prove to be popular with audiences looking for distraction and escapism.[22] Plans to release *The Nazi Strike* and others in the series were canceled, although the films did circulate within military training camps and in war plants as incentive films.

The story of German geopolitics even inspired a short film made by MGM in 1943 as a "victory short," a category of twenty-minute films coordinated between the studios and the state as crash-course instruction on wartime issues.[23] Entitled *Plan for Destruction* (1943) and directed by Edward Cahn, the short dramatized the story of Karl Haushofer. Structured around a wise and worldly American narrator pictured at a desk, it begins with Haushofer's humiliation after the defeat of World War I and his determination to make Germany great again. As a professor of geography at Munich, Haushofer is shown contemplating the globe in his study and instructing students in geopolitics in a large lecture theater. A tall, clean-cut actor overacts the part of Rudolf Hess, and we are shown Hess introducing Haushofer to Hitler (with his back to the camera) in Landsberg Prison as he writes *Mein Kampf* (see Figure 7). The film proceeds to inform us that with Hitler's ascent to power, a Geopolitical Institute at the University of Munich is established with unlimited funds and more than one thousand employees, all under the control of Haushofer, who reports directly to Hitler. This Geopolitical Institute is envisioned as a central receiving area for Nazi espionage across the globe, by spies disguised as tourists, by exchange students and athletes, or by Germans permanently living abroad. Letters, parcels, and secret deliveries in diplomatic pouches pour in from all over the world, while an army of researchers, amid card catalogs and observational equipment (magnifying glasses and microscopes), study information— "information of such scope and detail as has never been possessed by any other aggressor in history" (see Figure 8). The screen version of Haushofer is pictured instructing military officers on global geography while we are informed that Haushofer's files provided the key to victory in the various Nazi invasions (see Figures 9 and 10). With news of these victories, Haushofer and his army colleagues break out their "best Rhine wine" to toast world domination (see Figure 11). Haushofer's plan is for Germany to master the heartland and link up with the Japanese so the hemisphere of the Americas will be confronted by two sets of pincers, the Japanese

Figure 7. The triangle of conspiracy. Hollywood's Haushofer (in middle) with Hess meets Hitler (back to the camera) in Landsburg prison. Movie still from Plan for Destruction. *Copyright 1943 Turner Entertainment Co. All rights reserved.*

from the east and the Germans from the west. "Haushofer's plan for human enslavement," the narrator concludes, "is unquestionably the most elaborate blueprint for destruction ever devised." Fortunately, the Allied nations have figured it out and are now "slashing the tentacles of the geopolitical octopus with a brand of geopolitics of our own." With this image of penetration thwarted, the film ends with a ringing declaration for American youth to avoid overconfidence and remain firm in the fight against the enemy.

The Middle-Brow Policy Narrative

Distinguishable from the crude and conspiratorial reasoning found in popular narrations of geopolitics are the more considered analyses of geopolitics that began appearing in 1941 and 1942. In a series of speeches, articles, and pamphlets and in a book, the German émigré professor Hans Weigert took issue with the popular reading of German geopolitics, particularly that of Sondern.[24] Weigert argued that Sondern's claims were "sensationalist" and distanced his analy-

Figure 8. The site of power. Hollywood's image of the nonexistent Geopolitical Institute. Here is gathered "information of such scope and detail as has never been possessed by any other aggressor in history." Movie still from Plan for Destruction. *Copyright* 1943 *Turner Entertainment Co. All rights reserved.*

sis from the "recent fashion of making Haushofer the mystery man behind the curtain, his Institute a nest of international spies, and geopolitics, a superman-science on Nazi soil."[25] One negative effect of this, he argued, is to overestimate the superiority of the enemy and allow a defeatist attitude to develop at home.[26] Nevertheless, Weigert does not fully repudiate the conspiratorial narrative, suggesting later in his argument that "if we strip Sondern's words of their glamour and mystery, he is right in speaking of the '1,000 scientists behind Hitler.' "[27]

The disavowal yet reinscription of the significance of German geopolitics in Nazi foreign policy is characteristic of the middle-brow policy narration of geopolitics. Its characteristic features are the rejection of the account of German geopolitics as a superscience and of Haushofer as the man behind Hitler and yet the claim that Haushofer exercises a strong influence over Hitler's thinking and that the study of German geopolitics not only provides privileged

Figure 9. Master of the global map. Haushofer instructs representatives of the German state in global geopolitics at the Geopolitical Institute. Movie still from Plan for Destruction. *Copyright 1943 Turner Entertainment Co. All rights reserved.*

insight into the thinking behind Nazi foreign policy but, purified of its Nazism, is also a powerful practice that Allies would do well to cultivate. Whereas the popular narrative's dominant form is one of drama and conspiracy, the middle-brow narrative form is given over to serious imperative policy pontification, a what-needs-to-be-done form that does not necessarily reject the poetics of the dramatic and the conspiratorial genre so much as give it a serious gloss and integrate it into its imperatives. Within this general structure, there are a series of competing theses on German geopolitics. The four I wish to examine here are those of Hans Weigert, Robert Strausz-Hupé, Andreas Dorpalen, and Derwent Whittlesey.

The significance of German geopolitics for Hans Weigert was that it represented a "workshop for army rule." Following Oswald Spengler's prediction in 1933 that armies, not parties, would be the future form of power, Weigert read Haushoferism as the weltanschauung of the army that would, sooner or later, clash with the

Figure 10. Plotting world domination. More geopolitics lessons from Herr Professor Haushofer. Movie still from Plan for Destruction. *Copyright 1943 Turner Entertainment Co. All rights reserved.*

racialist ideology of Hitler and the Nazi Party. Instead of conceiving of Haushofer as the "man behind Hitler," Weigert argued that Haushofer could well be the "man after Hitler" in that he laid the groundwork for a new weltanschauung, "which might prove strong enough to replace Nazi ideology as a mass religion."[28] German geopolitics, Weigert suggested, "might be seen as the spiritual equipment for an army ideology which is on its way to overcome the primitive doctrines of Hitlerism."[29]

The subsequent flight of Hess to Scotland in the spring of 1941 was interpreted by Weigert as further evidence for his thesis. "Haushofer's geopolitics tried hard, but tried in vain, to win over the Party to its grand strategy." Hess's flight reflects symbolically the break between Haushoferism and Hitlerism.[30] The invasion of Russia by Hitler was another mark of the break, with Hitler deliberately ignoring the advice of his would-be mentor.

In contrast to Weigert's thesis, Strausz-Hupé's narration of German geopolitics has no distinction between Haushoferism and Hit-

Figure 11. Breaking out the best Rhine wine. Haushofer and the German military toast to world domination. Movie still from Plan for Destruction. *Copyright 1943 Turner Entertainment Co. All rights reserved.*

lerism, although Strausz-Hupé does note the detachment of Haushofer's writings from the Nazi party line.[31] *Geopolitik,* for Strausz-Hupé, is not a discrete army ideology but a pervasive revolutionary spatial doctrine that has "deeply influenced not only the thought processes of the leading Nazis but all branches of German public life."[32] Strausz-Hupé dismisses Weigert's thesis by suggesting that the army in Germany, just as in other totalitarian countries, is at the disposal of the totalitarian leader. It is subservient to the Nazi Party "because it lacks an ideology which could attract the German masses. It is a monstrous machine, it has no will of its own."[33] At certain points, Strausz-Hupé distances himself from the view that Haushofer's writings contain a master plan for German foreign policy and from exaggerations of Haushofer's influence over Hitler.[34] Nevertheless, his narration is notably more conspiratorial than that of Weigert. Geopolitics camouflages its significance with obscure language and obtuse terminology. Crafty Germans deliberately depreciate the importance of the new science while foreign observers are thrown off its scent by sly hints that geopolitics was not to be taken seriously.[35] Strausz-Hupé, however, sniffs a hidden truth behind the verbiage and confusion:

> A good deal of confusion as to the aims of Nazi leadership could have been averted by seeking the key to Nazi policies in their preoccupation with space. At the end of the Nazi rainbow lies not a new kind of economics or the eugenically perfect state, but physically dominated space. To this end Nazi Germany has bent her will and all her science. (119)

Strausz-Hupé's narration thus works by ostensibly downplaying the personal significance of Haushofer but nevertheless holding that geopolitics is the master plan or blueprint for Nazi world conquest.[36] Haushofer's teachings were intended, he argued, not as a fixed plan of action, but as a set of principles that would enable Germany's rulers always to choose the right course (71). The "key to Hitler's global mind is German geopolitics" (vii). Geopolitics is the system of thought that unlocks the mystery of Nazi foreign policy and reveals its supposed rationality. At base, however, this rationality is merely a moribund German craving for world power (x).

One distinguishing characteristic of Strausz-Hupé's narration of geopolitics is its refusal to associate Haushofer and German geopolitics with any particular foreign policy position. Unlike Weigert, therefore, Strausz-Hupé does not read the Nazi-Soviet pact as a triumph for Haushofer and the invasion of Russia as his fall from grace. The validity of Haushofer's theories, he suggests, is unaffected by the manner in which Germany secures domination of the "Heartland" (79). Strausz-Hupé's narrative, therefore, is not one that sets out to explore the influence of *Geopolitik* on Nazi foreign policy. He treats Haushofer's *Geopolitik* and Nazi foreign policy as a homogeneous monolith: "There is no reason to believe that Hitler consulted Haushofer when making his momentous decisions. He did not need to. *Geopolitik is* Nazi foreign policy" (79).

It is important to note, however, that *Geopolitik* for Strausz-Hupé is a degenerate form of geopolitics, a fallen geopolitics. Although his terminology is not always consistent and his arguments are hyperbolic, Strauss-Hupé clearly believes that the geopolitical approach to international politics is an important one. The universal truths of geopolitics are that space is power and that international politics is a struggle between different states for space. Various intellectuals throughout history, from Montesquieu, Alexander Hamilton, and von Bulow to Frederich List, Ratzel, and Mackinder, appreciated this. The teachings of Karl Haushofer are a German codification of the objective truths of geopolitics and have thus produced powerful insights into the course of global affairs. However, *Geopolitik* has allowed itself to be detached from its solid foundations. *Geopolitik* has been taken over by the Nazi ideologists. "Under political pressure *Geopolitik* slipped, one by one, the moorings which had held it to the firm ground of political geography and drifted into the shifting currents of Nazi propaganda" (26). Solid truths gave way to a tide of ideology.

While the majority of narrations of German geopolitics at the time locate its origins in European geographical and political thought, the account of Andreas Dorpalen stresses the influence of Haushofer's time in Japan on his thinking and on the development of *Geopolitik*. The success of Japanese foreign policy in 1910 in annexing the Korean peninsula and the programmatic vision of Foreign Minister Count Komura impressed Haushofer deeply, according to Dorpalen.[37] Their supposed skill at analyzing and interpreting the complex mix of factors that go into the dynamic functioning of world politics enabled them to evaluate and detect the point of least resistance to their goals. After careful preparation, they struck against this point and realized their ambitions with the minimum of risk to their country. This, in Dorpalen's interpretation, was geopolitics at its best to Haushofer.

Like the other middle-brow analysts, Dorpalen makes a distinction between geopolitics as a method and the use to which it can be put. "If applied to the realization of legitimate objectives, it is fully acceptable" (14). In Japanese and German hands, however, it has been perverted and prostituted in a drive for domination. Geopolitics, therefore, is a policy of conquest. After Germany's defeat in World War I, Haushofer set himself the task of schooling Germany geopolitically. He lost no time in spreading his ideas through lectures, maps, charts, books, and the journal *Zeitschrift für Geopolitik*. "By the time Hitler rose to power, all Germany was united in its demand for living space" (17). At the Geopolitical Institute in Munich, Germany's future leaders received the "world-political" training that the leaders of the Kaiser's Germany were so sadly lacking. Leaders

were taught to think "in continents." Germany's struggle is presented
as part of a huge struggle across the entire Eurasian continent against
the hegemony of the Western democracies. The Geopolitical Institute, Dorpalen concludes, maps the blueprints for world conquest
by Germany.

Like Strausz-Hupé, Haushofer's *Geopolitik* and Nazi foreign policy are treated as virtually identical, the former providing the ideas
for the conduct of the latter. Geopolitical concepts, Dorpalen claims,
have by and large shaped German foreign policy ever since the Nazis
came to power in 1933 (18). Nevertheless, he does identify a chasm
between the uncultured elements of the Nazi Party, with their racialist doctrines, and the cultured professor, who "holds on to the firm
belief that space rather than race is the touchstone of mankind's des-
tiny" (20). Haushofer's influence over Hitler is indirect and is exercised "through his disciples who, trained to see the world through
his eyes, translate his ideas into practical politics" (20). Dorpalen
hedges on Haushofer and Barbarossa, suggesting that Haushofer
must have viewed it with misgivings but that it is not entirely impossible that he did advocate the attack on the Soviet Union (155).
In keeping with the style of his whole work, Dorpalen also detects
Haushofer's hand in Hess's flight to Scotland.

Derwent Whittlesey's *German Strategy of World Conquest* is a
remarkably blunt analysis of German geopolitics and German political life. It begins with the bald claim that Germany intended to
dominate the world and had a plan for doing so. "To seize the earth
bit by bit, no matter what the order of the taking, is Germany's working formula for domination of the world."[38] For a thousand years,
according to Whittlesey, Germans had been pushing east from their
Central European homeland. Prussian and later German statecraft
adhered to the rules laid down in the *Political Testament* of Frederick the Great of deliberately provoked wars, piecemeal territorial conquest, and annexation of territory. Modern Germany used speculative and scientific thought to justify its goal of world conquest. Kant's
speculative philosophy "provided a cogent philosophical justification" for German military aggressiveness, as did the philosophies
of Schlegel, Hegel, and Nietzsche. Economics, psychology, and geography have all been used by the German state to further its global
ambitions. The general staff of the German state became a clearing-
house for all available information about the world's natural re-

sources, its officers casting "a practiced eye over the economic and political geography" of foreign states. Whittlesey documents the "plans of 1895 and 1911," which were specific Pan-German blueprints for rearranging the political map in Europe. The gathering of exact geographical information by the Pan-German movement foreshadowed German geopolitics, an intellectual plant nourished by German militarist thought, war, and defeat after World War I.

Geopolitics, for Whittlesey, is not a contemporary creation of Karl Haushofer but the product of a national habit of mind that reaches far back into German history (70). Geopolitics is not a rigorous discipline but a mixture of science with political aspiration, a hybrid philosophy that claims to pursue the scientific study of political geography to further the state. In fact, Whittlesey argues, geopolitics is merely a propaganda vehicle for German territorial aspirations. Anything goes in the utilization of geography in the interests of political expediency. The Institute of Geopolitics at the University of Munich has a research staff of more than eighty specialists who work with similar organizations on fact-finding for future space planning by the German state. Haushofer's twofold career as a brilliant staff man in the army and as the leader of thought on geopolitics is illustrative of the close ties between the German general staff and geopolitics. Significantly, prominent geopoliticians hold commissions in the army and representatives of the general staff are found throughout the civil administration of Germany (113–15).

That Germany is an inherently expansionist power with deep and ancient expansionist desires is the level of Whittlesey's argument. His analysis lacks any discussion of politics, political economy, or an analysis of the Nazi state. Germany simply is this way, a creator and devotee of planned territorial expansionism. "Intermittent and unorganized forays by German feudal lords against neighboring Slavs a thousand years ago have been replayed by a comprehensive and detailed design for world conquest" (258). "The proposed German substitute for the present political structure of the earth is unified political control, with Germany in charge" (261). As an exposé of propaganda, Whittlesey's book is also an extraordinary example of the practice, a geopolitics in the name of "political geography" against geopolitics, a mix of positivism and ideology against positivism and ideology, a book appropriately stamped with the slogan "Books are weapons in the war of ideas."[39]

"LET US LEARN OUR GEOPOLITICS"

A Pervasive Experience of Lack

In the various narrations of German geopolitics, it is common to find reference to Haushofer's frequent quotation of Ovid's maxim "Fas est ab hoste doceri" (It is one's duty to learn from the enemy).[40] The maxim is noted to illustrate how Haushofer read Mackinder's ideas about the importance of the heartland and how he adapted them to his own ends. The maxim, however, takes on an additional meaning in the context of its use, for the very project motivating the narrations of German geopolitics in U.S. political discourse during World War II was precisely to learn from the Nazi enemy, from Haushofer and his methods of geopolitical thought. A common theme running throughout narrations of German geopolitics is a pervasive sense of lack, a feeling that the Allies are missing the hardheaded, potent power of geopolitics. In the most direct psychoanalytic terms, the Germans have it and we, the United States, do not. As a consequence the United States is poorly equipped to deal with threats and aggression. Weigert put it as follows:

> The lack of centers where American students and soldiers can, like the German youth in Munich, be trained to understand the facts and to think in terms of political geography and geopolitics seems to me to be a regrettable flaw in the endeavor to organize democracies against the totalitarian onslaught.[41]

Like so many others, Weigert concludes his reading of German geopolitics with the admonition that "we must learn our own geopolitics."[42] This is also the theme of a fellow German exile, Frederick Schuman, who identifies an "inner essence" or "hard core of worldly wisdom" within geopolitics that is to be distinguished from the mysticism and ideological rationalization of imperialism found in Haushofer's brand of geopolitics.[43] Schuman understands this wisdom in quasi-religious terms: geopoliticians are apostles who preach a strategic gospel. He argues that in a "world of rival sovereignties the art of combining geography and strategy on a world scale is indispensable to the leaders of every nation that seeks to win or keep a world position." If the Allies are to win, they will need their own geopolitical centers and "their own counterpart of Haushofer, stripped of its cynicism and mythology."[44] In so defining the issue in terms of geopolitical education, centers, and father figures, however, Schuman was himself (re)creating a mythology for geopolitics. Let us look at these three themes in greater detail.

The Site of Power

In his 1941 book *The Time Is Now!* Pierre Van Passen claims that although the Geopolitical Institute was only discovered by America in 1939, it has existed since 1897 and its ideology has fermented the minds of German political thinkers since 1870. We are, he argues, four decades late in recognizing the fact that the German government "enthusiastically sponsored a planning academy whose function was to develop a long-range project for the domination of the world."[45] According to Werner Cahnman, Haushofer keeps in his Geopolitical Institute "a file on almost everything and everybody in every country and in every part of every country on the face of the globe."[46]

It is indeed puzzling that such envisionings of a nonexistent Geopolitical Institute should recur again and again in U.S. accounts of German geopolitics during World War II. Haushofer did, of course, preside over the German Academy, but this organization cannot be compared to the institute described in numerous accounts of geopolitics at this time. What then accounts for this particular myth about German geopolitics? One possible explanation is to consider discourse on German geopolitics within a larger tradition of discourses of danger on foreign threats throughout U.S. history. Within these discourses of danger, there is often an institution or organization that is considered to be the seat of planning and conspiracy against the United States. Commonly draped in mystery, this site of power is represented as being the foreign power's headquarters of rationality, the secretive coordinating place where blueprints for world conquest are schemed and studied, instigated, and implemented. In the early part of the twentieth century, the Prussian general staff occupied such a position in discourse on Germany. The negative terms of the late-nineteenth-century discourse on Prussian militarism and Junkerism transposed well onto the Third Reich and Nazism, terms that resonated with the experience of many in the German émigré communities who came to the United States to escape bureaucratic militarism at home. Similar discourses were produced about the Japanese general staff and state bureaucrats. During the subsequent Cold War, the Kremlin became the enemy site of power for many cold warriors, the place that coordinated the worldwide Communist conspiracy. In certain contemporary geoeconomic discourses in the United States, Japan's Ministry of International Trade and Industry (MITI) occupies a similar position.

There is a strong case for considering the myth of a German geopolitical institute in such terms. Sondern's vivid description of a vast strategic index expresses a fantasy of an ultimate global panopticon, an all-seeing, all-knowing institution that gathers every piece of information, no matter how trivial, and feeds it into a rationality machine that then uses this information to implement some diabolical global design. Even the less conspiratorial middle-brow narratives have references to the Geopolitical Institute at Munich under the personal direction of Haushofer with a large research staff.[47] Strausz-Hupé represents Munich University as a laboratory of world political ideas. The image of a laboratory and the frequent casting of Haushofer as a brain trust or superbrain behind the Nazi design for world conquest reiterates a common movie-informed fantasy about a crazy scientist (usually played by a foreign actor like Boris Karloff or Peter Lorre in Hollywood movies) with a diabolical design for the world.

Yet, representations of the site of power are not unambiguously dystopic, for the dream of a utopian place where information can be collected and analyzed, blueprints devised, and geopolitical calculations made with precision is evident even in the condemnation of German geopolitics and its Geopolitical Institute.[48] Even before America's entry into the war, a campaign to establish an American Institute of Geopolitics along the lines of Haushofer's imaginary institute had begun. In 1940 Eric Archdeacon wrote to Sloan Colt, the president of Bankers Trust Company in New York, urging business leaders to take the lead in the creation of such an institute. Since business had developed economic forecasting as a science, why not apply the science of forecasting to international relations and politics?[49] By March 1942 Archdeacon had elaborated his suggestion into a fully developed proposal for an independent American Institute of Political Geography to be headquartered in Washington, D.C. The proposed institute would serve as "the 20th century fact-finding agency for executive government which replaces to a substantial degree guesswork and methods of trial and error in national and international government executive work."[50] Archdeacon's proposal for an open civilian and military institute was opposed by a number of military figures, including West Point's Colonel Herman Beukema, who expressed reservations about giving civilians secret and confidential information. Nevertheless it found support among elements of the business elite. *Business Week* editorialized that business leaders could not afford to overlook the "growing Washington interest in the new science of geopolitics."[51] In a second editorial, "The Case for Geopolitics," the magazine speculated on the possible creation of an American Institute of Geopolitics, noting that U.S. leadership of the war "demands that we define our long-term needs and objectives, tighten up all our planning organizations, and objectively coordinate all of our activities."[52]

As a consequence of the arguments of Archdeacon, Beukema, and others, the War Department established in June 1942 a Geopolitical Section within the Military Intelligence Service. Its stated objectives were "to study physical, economic, political, and ethnological geography in order to advise on measures of national security and assurance of continued peace in the post-war world."[53] Lieutenant William S. Culbertson, the driving force within the section, had a vision of its role that was similar to that of Archdeacon. He helped organize a business roundtable on geopolitics in New York City in September 1942 and numerous academic conferences addressing geopolitics and, in October 1942, persuaded Henry Luce to convene a meeting of top publishers and newspaper figures on the topic of geopolitics.[54] Furthermore, he developed an elaborate consultant role for the nation's top geopolitical intellectuals, like Edward Mead Earle, Harold Sprout, Nicholas Spykman, Father Edmund Walsh, and Derwent Whittlesey. Culbertson's extension of the section's activities beyond his original mandate provoked the ire of his superiors after a while, and the Geopolitical Section was abolished in 1943. What remained, however, was the dream of a global panopticon of intelligence, a desire the Office of Strategic Services had already begun to institutionalize and a desire that would later propel the establishment of the Central Intelligence Agency and the National Security Agency.

The Power of Sight

A second aspect of the appeal of geopolitics in U.S. discourse at this time concerns the faculties and powers usually attributed to geopolitics and those who practice it. There are many versions of these powers, but they usually come back to the power of sight and insight. Cahnman's account of the methods of geopolitics notes that geopolitics differs from political geography because of the extent to which it is a predictive science. Prediction is not mere forecast but "vision as a synoptic view born of knowledge plus insight."[55] Seeing synoptically is seeing things from a holistic point of view, an apprehension of the whole made possible by a knowledge of the parts, and vice versa. Whittlesey describes Haushofer as "the oracle."[56] Geopolitics, for Weigert, is about envisaging the world in global terms, about understanding the world as a closed unit. The dynamic focus of geopolitics on the earth and space leads to a "new way of envisaging man's role on earth," a method by which "the future can be predicted by interpreting the fateful signs revealed in the pattern of the earth."[57] In his article on maps, he describes the suggestive, dynamic maps of German geopolitics as the work of "geopolitical magicians."

This equating of geopolitics with magic is also made by Hans Speier in his discussion of German cartography.[58] Geopolitics, in Weigert's account, is a form of prognostication, a type of geomancy. For example, he claims that Haushofer and his circle "foresaw" the doom of France, recognized the significance of Charles de Gaulle when he was little known in his own country, and foresaw Pearl Harbor. The two Haushofers (Karl and Albrecht) are oracles of power politics; the "amazing accuracy" of the Haushofer forecasts "must not, however, induce us to ascribe supernatural powers to Germany's 'secret' science."[59] The very construction of this sentence reveals the paradoxical quality of the narration of German geopolitics. Geopolitics is not a supernatural science, yet its practitioners can predict with amazing accuracy. Geopolitics is not beyond the realm of the rational, yet its use can produce analyses of global politics that go beyond the everyday operation of reason in international politics. Geopolitics, in short, is set up as an uncanny form of rationality that can yield remarkable results. It is a method that, if cultivated with sufficient resources, energy, and dedication, can bestow upon its practitioners a power beyond the ordinary. Geopolitics is insight, the figure of the geopolitician a seer. This latter term captures the ambiguous rationality of the geopolitician; a scientist yet also a prophet, a positivist yet also a creative, envisioning artist.[60]

gift but one "possessed at all times by men whose genius carried them beyond their epochs."[61] The identity "genius" is a central part of Weigert's reading of geopolitics. Mackinder and Haushofer, for him, are geniuses, the former, in his opinion, the greatest geographer that ever lived.[62]

This narration of geopolitics as an out-of-the-ordinary visual capability is also to be found in Strausz-Hupé. Haushofer, as the "master" of geopolitics, is represented as a man with a mission: "to open the eyes of his people to the full meaning of space."[62] Haushofer's "organizational genius and clever showmanship opened the eyes of the Nazi party leaders to the potentialities of *Geopolitik* and at the same time carried its concepts to the German people" (48). He regarded himself as the "preceptor" of the German statesman (73). Haushofer's books all contain "brilliant insights," yet, according to Strausz-Hupé, they are also padded with "unbelievable platitudes" (75).

Enlightenment

A third dimension of the appeal of geopolitics is its association with enlightenment through proper training. This dimension of geopolitics had already been articulated before the war. In one of the earliest articles on geopolitics outside the specialist field of political geography, Ewald Schnitzer, writing in the *Harvard Educational Review,* argues that geopolitics is a vital and necessary part of any "social education," education for citizenship.[64] Geopolitics, he argues, is useful not simply in the "civic training of the masses." It also has a vital role in training the political and economic leaders of any nation. He notes that almost all periods of successful political administration utilized geographical knowledge. "One-sided juristic thinking has been responsible for many of the misconceptions and ill-adjustments in our political world, because it preferred 'to put the state on paper, instead of on solid ground' (Kjellen). It is the goal of geopolitics to remove this shortcoming in the training of state functionaries as well as in the education of the people towards a more enlightened citizenship" (508). Among the benefits of using geopolitics in social education suggested by Schnitzer is its facilitating fuller discussion of "inheritance and racial hygiene" (516).

Whittlesey's argument champions a geopolitics in the name of political geography. To maintain peaceful neighborliness in world politics will "take an understanding of the earth-base of human society at least equal to that possessed by the better informed Germans."[65] This knowledge, he suggests, can be obtained and disseminated "only by a long-term program of learning about the earth in which everybody will participate." A vast increase in factual information about all parts of the earth is needed, as are more detailed maps and public information agencies to disseminate knowledge of the earth. That information and knowledge might involve interpretative decisions and political power does not occur to Whittlesey, so complete is his faith in progress through the positivist accumulation of facts. "Grasp of the elements essential to a stable world can come only with sure knowledge of the whole earth" (268).

That what Whittlesey was calling for was not that different from German geopolitics is illustrated by Weigert's account of its positive educational aspects. In his pamphlet *German Geopolitics,* for example, Weigert declares:

> The multitude of surveys of geopolitical and geographical subjects all over the world and the quality of these studies have made, in not much more than fifteen years, the Munich school a geopolitical center which has no competition.... The enthusiasm in the field of geopolitics which Haushofer kindled has had an enormous educational result. Hundreds of students...and dozens of young officers of the Munich garrison formed ranks of Haushofer's followers to become educated and trained in careful, industrious geopolitical work. It cannot be emphasized how great was the educational achievement of this man at a time when the average youth, disillusioned, turned to political phrases and slogans.[66]

What is interesting about this, besides the emphasis on education and training, is that Haushofer is represented as a model modernist man committed to education, hard work, and social progress. Haushofer's project is an enlightenment project, a success story undertaken in opposition to degenerate tendencies in modernity, a project built upon the notion of social progress through disciplined training and learning. This aspect of German geopolitics is particularly appealing to public policy intellectuals. Haushofer's story becomes a morality story of how we can be better, how we need to address the slackness, the "softheadedness," and the ignorance and indifference to world affairs in our own society.

THE MYTHOLOGY OF GEOPOLITICS AND THE
DIALECTIC OF ENLIGHTENMENT

While German geopolitics was being narrated by émigré public policy intellectuals like Dorpalen, Schuman, Weigert, and Strausz-Hupé, who were carving out a space for a good (German) geopolitics, two other German intellectuals who had fled Nazi persecution were working on a book that they would eventually finish in Los Angeles in 1944. Max Horkheimer and Theodor Adorno's *Dialectic of Enlightenment* is known today as a classic statement of what only later became known as the Frankfurt School. The book is a deeply pessimistic consideration of why the process of enlightenment has brought with it so much destruction and barbarism. Although the sweeping nature of some of the book's claims and statements are clearly problematic, Horkheimer and Adorno's thesis offers an interesting counterperspective on the story of German geopolitics in U.S. political discourse. The Enlightenment tradition of progressive thought, they claim, has always aimed at liberating men from fear and establishing their sovereignty. "The program of the Enlightenment is premised on the disenchantment of the world, on the dissolution of myths and the substitution of knowledge for fancy."[67] The Enlightenment and mythology, therefore, are supposedly antithetical forms of knowledge.

What Horkheimer and Adorno point out, however, is that "myths always realize enlightenment and enlightenment with every step becomes more deeply involved in mythology. It receives all its matter from the myths, in order to destroy them; and even as a judge it comes under the mythic curse" (11–12). In other words, the Enlightenment itself operates as a mythic system, a system that founds itself on the dissolution of the religious and the magical yet requires the very principles of such mythic systems to construct its narrations and motivate its imperatives. Horkheimer and Adorno argue that the Enlightenment's program of domination is a secularized version of the religious belief that God controlled the world. The human subject confronts the natural object as an inferior, external other.[68] Their identification of the persistence of the mythical in the discourse of the Enlightenment is, I believe, a suggestive means of approaching the case of German geopolitics in U.S. political discourse, for it allows us to sketch out the often paradoxical nature of the emergent mythology of geopolitics at this time. I wish to conclude by considering geopolitics in these terms.

Enlightening Myths

It is already well established that certain elements in the representation of German geopolitics in U.S. political discourse during World War II are mythical and fictitious. Less well established but perhaps more apposite is the argument that all the representations of German geopolitics at this time, both popular and middle-brow policy narratives, functioned positively as enlightening myths about Nazi Germany. Such representations held out the promise of making Nazi foreign policy rational and comprehensible. The study of German geopolitics promised, at a general level, access to the "secret" Nazi blueprint for world conquest. Geopolitics, therefore, was an important element in making Nazism meaningful to millions of Americans. It offered a quick and easily understood explanation of Nazi foreign policy. It helped deal with the challenge Nazism posed to the world of reason and enlightenment in general, the challenge to the very nature of modernity itself, by attributing certain principles of reason, albeit misguided and frequently described as mystical and cynical, to the foreign policy of the Nazi state.

While one could argue that this had a certain propaganda value, one could also argue that these narrations of German geopolitics promoted a fatal misreading of the Nazi state in U.S. political culture. On a general philosophical level, there was a failure to think through the crisis in state-led modernity and Western civilization that Nazism represented (a philosophical exercise that has arguably still not taken place). That one could isolate and analyze Nazism in a rational manner using a rational interpretative framework (which includes the categories "madness" and "mysticism") was seriously in question. On a more historical level, readings of German geopolitics clearly misread the way Nazi foreign policy functioned and attributed a rationality to it that it did not have. The persistent stress on bureaucratic blueprints for world conquest seriously misrepresented the haphazard and personality-driven method of policymaking (although not policy implementation) in Nazi Germany. Also, *Geopolitik* was never the coherent foreign policy philosophy it was made out to be. Nor could the relationship between the German army and Nazism be reduced to formulaic terms or abstract Spenglerian theses. The consequence of this was not simply to exaggerate the strength and coherence of Nazi foreign policy. Perhaps the most fatal consequence was the complete underestimation of the power of racialist ideologies in the functioning of the Nazi power structure. Readings of German geopolitics marginalized anti-Semitism and the foremost diabolical possibility that the Nazis did attempt: the complete annihilation of the Jews of Europe. If German geopolitics was a myth that enlightened, it was also a myth that blinded.

Mythical Enlightenment

One could argue that one mark of the failure to think through the implications of Nazism for a state-guided modernity was the mythical enlightenment that commentators drew from German geopolitics for the theory and practice of U.S. foreign policy. In expressing the desire to overcome a lack, to establish sites of power, cultivate the power of sight, and foster enlightened geographical consciousness throughout the nation, U.S. narrators of German geopolitics were drawing upon the same imaginary economy as that appealed to in German geopolitics. The U.S. state was gearing itself for world power, and geopolitics came to be understood as essential to that preparation. The identity of geopolitics-as-object, as it came to be mirrored in U.S. political discourse, was no less reliant on mysticism and magic, the discourse of prophecy and uncanny insight, than its ostensible opposite in Germany. Geopolitics both enlightened and mythologized the earth; it promised the disenchantment of the surface of international affairs by reenchanting that very surface with tales of undisclosed strategic space and magical formulas that compelled the earth to reveal its secret strategic pathways and faultlines. Its intellectuals were positivists yet also prophets, scientists yet also seers, scholarly monks of the terrestrial who studied the textuality of continents and the saints of the geographical tradition in order to preach, in the name of geopolitical fathers and prophets before them, the laws of the earth.

It would be misleading to accept the assumption, too often made in accounts of German geopolitics, that the U.S. state and U.S. strategic culture did not think in geopolitical terms before World War II. U.S. history, as Strausz-Hupé and many others argued at this time (following, incidentally, the arguments of Haushofer and other contributors to the *Zeitschrift für Geopolitik*), can easily be read in geopolitical terms. What was important was to assert a *globalist* geopolitics to counter the *isolationist* geopolitics of U.S. political culture. The U.S. public needed to think in terms of continents and global spaces, not parochial regions and separate hemispheres. In making the case for geopolitics as a new positive conceptual practice in U.S. strategic culture, its émigré authorities of delimitation were grafting the term onto the preexistent landscape of foreign-policy analysis as another justification for an interventionist U.S. foreign policy.[69] Geopolitics did not have any essential meaning in and of itself. Its symbolic grid of specification was the discourse of Nazi danger and the imperative of responding to this Nazi threat militarily. At this time, geopolitics did not necessarily mean political realism. Weigert's call for a "geopolitics of peace" was explicitly antirealist, while other instances of geopolitics, such as the clumsy effort by George Renner to draw new political maps of the world's continents,[70] were far from that which could be described as "realist."[71]

What was rarely problematized, as the concept of geopolitics emerged in America, was the very possibility of a rational geopolitical knowledge and method, the very possibility of state-sponsored enlightenment. In the narrations of German geopolitics, the principle of the Enlightenment (and mythologization) was always reaffirmed. This point is not trivial, for the type of enlightened reason championed in these narrations was what Horkheimer and Adorno identified as instrumental reason, a debased, one-dimensional form of reason used to shore up systems of power and identity. U.S. narrations of German geopolitics during World War II saved geopolitics from itself and resecured its functioning as an instrumental form of reasoning. In making geopolitics-as-object, these narrations expressed an imaginary that was soon channeled by the symbolics of the Cold War. Empowered in this process was a new émigré policy priesthood (of which Robert Strausz-Hupé was a leading member), whose faith in the doctrine of geopolitics, to paraphrase Horkheimer and Adorno,[72] turned it into an instrument of rational administration by the wholly enlightened as they steered postwar society toward new world wars, both hot (in Korea and Vietnam) and cold (fighting a worldwide Communist conspiracy).[73]

5

Critical Approaches to "Geopolitics"

There is a will to essentiality which one should mistrust.

— MICHEL FOUCAULT
ANSWERING QUESTIONS ON GEOGRAPHY[1]

The emergence of the sign "geopolitics" in the first half of the twentieth century was a deeply contentious one, fraught with political consequences of varying kinds. After the shock of military defeat and the humiliation of the dictated peace of Versailles, the Weimar Republic proved to be fertile ground for the growth of a distinct German geopolitics. Geopolitical writings, in the words of one critic, "shot up like mushrooms after a summer rain."[2] That critic was Karl Wittfogel, a German Communist who in 1929 wrote the first systematic critique of the practice materializing at that time around the name "geopolitics." Since then geopolitics has inspired a variety of critical readings that have sought to isolate, critique, and, in certain cases, dismiss it as an illegitimate form of knowledge. From the orthodox Marxism of Wittfogel to the poststructuralism of contemporary critics, geopolitics throughout the twentieth century has provoked strong intellectual opposition and stimulated the production of "antigeopolitical" forms of knowledge.

The history of these various attempts to specify and critique geopolitics is of interest for a number of reasons. First, each of these critiques provides a window into the con-textuality of geopolitics as

141

a theoretical and political practice. The critical reading of geopolitics is more than simply a reading of an already existent object or immanent social phenomenon. Critical readings of geopolitics are ways in which geopolitics as a conceptual object is written, ideological inscriptions that assign a certain identity and coherence to it as part of an argument about its nature and relationship to state and society. Critical readings of geopolitics are significant moments in the writing of a truth around "geopolitics" and about geo-politics. They are thus important occasions for exploring how the concept of geopolitics is formed and how it functions. Second, critiques of geopolitics do not transcend the operation of networks of power/knowledge. The specification of geopolitics as a "pseudoscience," for example (a common practice in both Western and Soviet geography after World War II), is part of a discursive legitimation of a system of proper "scientific" geographical reasoning, together with an ensemble of institutions and intellectuals that are held to produce this "scientific" geography. Critical readings of geopolitics are themselves enmeshed in relations of power and can congeal geography and politics (geopolitics) into new forms of geo-power. As already demonstrated in the previous chapter, certain critical approaches to geopolitics have an ambivalent relationship to geopolitics as an intellectual activity. Many intellectuals could simultaneously be described as critical of geopolitics yet also as geopoliticians, figures who define their intellectual politics in opposition to "geopolitics" yet nevertheless work within the conceptual infrastructures that make geopolitics possible. Prominent intellectuals like Robert Strausz-Hupé, Karl Wittfogel, Isaiah Bowman, and others are ambiguous figures who traverse both the critique and practice of geopolitics, who are both against yet also for geopolitics.

To explore the meaning and implications of these contextual instabilities, indeterminacies, and ironies, this chapter reviews a series of attempts to isolate and critique geopolitics in a critical way. The various readings and writings of geopolitics considered in this chapter offer us a means to think about, first, the general problematic of geographical reasoning in global politics; second, the exorbitant meanings of geopolitics; and, third, the limits of the project of critical geopolitics. Out of many possible examples of critical approaches to geopolitics, I have chosen to focus on the accounts offered by Karl Wittfogel, Isaiah Bowman (and to a lesser extent Richard Hartshorne),

Yves Lacoste, Richard Ashley, and Simon Dalby, because each presents inscriptions of geopolitics that are distinctly different yet that reveal certain common difficulties. Although the context of these write-ups of the meaning of geopolitics are dissimilar, there are three common themes in my account of how each interprets geopolitics.

The first is that any critical theory of geopolitics needs to recognize that the problematic of geopolitics is a discursive, con-textual one that inevitably forces one to address questions concerning the politics of signification, the interpretative politics of reading and writing. Second, following on from this, critical theories of geopolitics must inevitably address the question of geopolitics as an ocularcentric system of knowledge, that is, a form of power/knowledge that relies on a Cartesian perspectivalism in order to function. Any critical interpretation of geopolitics must address the problematic of sight/site/cite as it operates in geographical reasoning as a whole. A failure to problematize the panopticonism that characterizes so much geographical discourse is ultimately a failure to problematize the conceptual infrastructures that make geopolitics possible. This is significant in that forms of knowledge that advertise themselves as antigeopolitical can and do continue to work within the very conceptual infrastructures that made geopolitics possible. Geopolitics and antigeopolitics are not necessarily opposites.

Finally, even if the question of geopolitics is recognized as a question of con-textuality, strategic power, and ocularcentrism, the particular specificity of the deployment and use of the sign geopolitics should be acknowledged and respected. Although the concept of geopolitics can be treated as a metaphor for the strategic operation of power in all forms of knowledge, this critical poststructuralist attitude is itself a strategy that effectively dissolves the particular historicity and spatiality of the deployment of geopolitics as an indeterminate but nevertheless congealed form of power/knowledge. In documenting the difficulties of critical approaches to geopolitics, I am making the case not for a general critique of geopolitics but for a careful genealogical approach to the problematic of the writing of global space by intellectuals of statecraft.

KARL WITTFOGEL: GEOPOLITICS AS BOURGEOIS IDEOLOGY

In 1929 Karl Wittfogel was emerging as one of a number of leading intellectuals in the German Communist Party (KPD). An active playwright in his youth, Wittfogel had been invited to join the newly established Frankfurt Institute for Social Research in 1925, four years after its founding by Felix Weil. At the time Wittfogel was the most politically active Communist member of the Institute, although he was a member because of his nonpolitical positivist work.[3] His overarching theoretical interest was in Marxism and the question of nature, an interest Wittfogel pursued regionally in publications on China. Later, after fleeing to the United States, renouncing his Marxism, and publicly smearing Owen Lattimore before a Senate committee in 1951, Wittfogel published *Oriental Despotism* (1957), the intellectual work for which he is most remembered.

Germany in 1929 was a country in a state of tremendous turmoil. Not only did the stock-market crash of that year contribute to deep depression in the global economy, but far-right extremists were rapidly gaining popular support among the unemployed, the natural constituency of the KPD. Karl Haushofer had founded the *Zeitschrift für Geopolitik* in 1924, and the romantic myths held by the conservative intellectuals associated with the journal were finding popular expression as Nazi street slogans. Geopolitics was also attracting the attention of certain leftist intellectuals, most notably the Social Democrat Georg Engelbert Graf and the British socialist James Francis Horrabin, both of whom suggested that Marxism failed to adequately address the question of nature in the determination of social life.

Wittfogel begins his critique "Geopolitics, Geographical Materialism, and Marxism" (1929) with an attack not on Haushofer or the political use of geographical myths by the Nazis, but on Graf's and Horrabin's suggestion that Marx did not adequately consider the question of nature. Since the Third (Communist) International in 1928, Social Democrats, not the Nazis, were considered the main enemy by the KPD and pejoratively labeled "Social Fascists."[4] Published in the journal *Under the Banner of Marxism* in three separate issues, the main purpose of Wittfogel's article was to demonstrate that the scattered remarks and concrete investigations of Marx and Engels together constitute a "coherent whole" that is the basis for a truly comprehensive, dialectical theory of nature and its relationship to the historical development of social formations. In the course of this necessary act of theoretical homage to the writings of

Marx and Engels, Wittfogel offered what can be considered a critical theory of geopolitics.

Geopolitics, Wittfogel claims at the outset, "represents an organic ideological complement to bourgeois-democratic practice" in Weimar Germany.[5] By suggesting that at least one-fourth of political life should be understood in terms of its earth-bound nature (the calculation is Haushofer's), geopolitics constitutes a regression to the old geographical materialist methods of the eighteenth and nineteenth centuries. This method, according to Wittfogel, postulated that geographical factors, whatever their character (climate, soil, location, physical terrain, even race), directly influence political life. The essential problem with this method is that it distorts the true reality that geographical factors do not directly influence but rather mediate the political sphere of life in human societies. In neglecting the necessary set of linkages connecting nature and the political domain, geopolitics ends as "crude distortion." It is scientifically "worthless" (23).

To substantiate his argument, Wittfogel provides a review of the works of the leading "epigones" of geographical materialism in the late nineteenth and early twentieth centuries. In Ferdinand Richthofen, a leading nineteenth-century German scholar on China, he finds a "combination of ineffectual geographical materialism with a completely unfounded eclecticism," which, he states, is repeated by all geopoliticians. Friedrich Ratzel's arguments, briefly noted already in chapter 1, are labeled absurd, "plainly misleading," "inaccurate," "completely false," "partial," obscure, and mystical (25–26). Ratzel's mistake was to let the economic sphere completely disappear from his analysis of the state's relationship with its "soil." The state is "not an earthworm." The population that constitutes the state-society totality does "not live directly from the *soil* but from the plants and animals that exist on the soil and are usually only produced and made consumable through *labor* (25; original emphasis). It is this neglect of the crucial link between nature and society that makes Ratzel's writings not a "system of *interlocking* scientific explanations but a conglomeration of mystifications *externally stuck together*" (26; original emphasis).

To Wittfogel, Rudolf Kjellen's empirical analyses of international politics at the beginning of the twentieth century retain the theoretical mysticism of Ratzel. But Kjellen wrote at a time when imperial-ism, according to Marxist-Leninist orthodoxy, had begun to take on a monopolistic-reactionary form. His new science of geopolitics "not only adapts to the needs of monopolistic-imperialistic capitalism," according to Wittfogel, but "even anticipates its future needs" (27). Wittfogel's reading of Karl Haushofer continues in this vein. Haus-hofer's statements are occasionally true, but he also falls back on the "old mystique" of arbitrary determinations. The geopolitics of the *Zeitschrift für Geopolitik* are ultimately completely uniform. They express the struggle of the bourgeoisie for the preservation of their privileges and the "necessity of pursuing an imperialist poli-tics for class objectives" (30).

Following consideration and criticism of the attempt to import the ideas of geopolitics into the labor movement by Graf and Horra-bin, Wittfogel turns to consider the eighteenth-century origins of geographical materialism. Geographical materialism, whether in its direct mechanistic French form or its more spiritualist German form, is reduced to an intellectual weapon of the bourgeois revolution. Its discovery of certain laws of nature undermined the persistent feu-dal-absolutist worldview, which was centered on God and notions of the divine. Behind Montesquieu's eighteenth-century materialism, Wittfogel suggests, stood the bourgeois demand for equality with the aristocracy and nobility. Geopolitics is merely the latest form of this old eighteenth-century geographical materialism. Whereas the pio-neers of geographical materialism sought to lay bare the dynamic laws of history with their method, their twentieth-century epigones have become more modest and only claim to be able to explain twenty-five percent of the truth.

Wittfogel's arguments against geopolitics as a scientific practice can be reduced to two fundamental claims. (1) Geopolitics, as a modern form of a mechanically determinist geographical material-ism, is a practice that *short-circuits* the objective levels of media-tion that separate nature in the raw from political life. The short-circuit method, as Wittfogel describes it, "designates a procedure (typical of the geopoliticians) which omits from the analysis one or more of the most important connecting links and thus leads to 'purely arbitrary determinations' which might occasionally be true but for the most part are only half-true or completely false because they are not in fact scientifically developed" (38). The concept of short-circuiting is a product of Wittfogel's faith in a positivist science of society. (2) As a bourgeois social science, geopolitics cannot articulate true general conceptions without, at the same time, articulating all the contradictions of monopolistic-imperialistic capitalism. Geopolitics is produced from the class perspective of the bourgeoisie. It exemplifies "the law of the diminishing power of perception of bourgeois social science." The most geopolitics can do is amass material, which the science of Marxism-Leninism must separate and reorganize into a completely new general conception (31).

Wittfogel's reliance on the metaphors of short-circuiting and diminishing perception reveal much about his epistemological commitments. The image of bourgeois science as producing arbitrary determinations normalizes Marxism as an objective science of the totality of human society. "Only Marx's conception connects social life with its real foundation, with the type of its material production" (53). Three key implications flow from these discursive claims. First, they grant absolute power to "Marxism" to divine the real (although the meaning of "Marxism" was always in contention). Marxism is made into a positivist, totalizing science. Its analysis of social life is holistic, objective, and correct, whereas other perspectives are partial, distorted, and arbitrary. It has the power to decide what is true and what is not. Although he had an ambiguous relationship with Stalin and was later to break from Communism completely, Wittfogel's totalizing scientific Marxism was not that radically different from the epistemology that sustained Stalinism.

Second, Wittfogel's Marxism is one that operates outside history and outside society. It is not sensitive to its own historical conditions of production and the limitations generated by this for the formation of sweeping generalizations about the whole of human history. Wittfogel's ideological rival in the Communist movement at this time, Georg Lukács, criticized Wittfogel's Marxism on just these grounds, arguing that Wittfogel lacked "a real critical and thus a real concrete analysis of phenomena."[6] One consequence of this failing is that Wittfogel conceptualizes the relationship between "Man" and "Nature" in a deeply gendered way. Taking up William Petty's remarks that labor is the father and earth the mother of material life, Wittfogel describes "the basic relationship" between "man" and "his natural environment" as one of father and mother, active movement and passive determination of direction (55).

Third, the type of intellectual practice sanctioned by Wittfogel's arguments is an extremely narrow one that privileges Marx, and to a lesser extent Engels, as the great readers of human history. Marx's texts are the Communist movement's bible, and the task of intellectuals is to practice an authoritative hermeneutics on these texts. The purpose of this textual exegesis is to get at what Marx really intended, to produce and reinforce an orthodoxy. Gary Ulmen claims that Wittfogel not only explained Marx's conception of historical materialism but goes beyond it.[7] Wittfogel did what Marx intended to do but did not do. Once his isolated remarks are gathered together, Marx's intent is held to reveal itself. This strategy of reading, however, is politically problematic, as are the details of Wittfogel's interpretation of Marx's writings.[8] To simplify a more involved argument, the assumption that texts have authoritative univocal meaning and that reading is revelation granted authoritarian power, at this time, to the central committees of Communist parties, all of whom were, by the late twenties, being forced to follow the theological line set by the former seminarian, Joseph Stalin.[9] Interpretative authoritarianism underpinned political authoritarianism.

The metaphor of seeing is extremely important to Wittfogel. Unlike bourgeois science, orthodox Marxism is a powerful form of seeing that renders things transparent. Bourgeois science produces mystifications. Orthodox Marxism produces clear truths. We can trace the operation of this powerful ocularcentrism throughout Wittfogel's text. The modern epigones of geographical materialism *reflect* the social contradictions of their time. Their *point of view* is that of the bourgeois. Wittfogel concedes that one can have "true partial insight" within "enlightenment" bourgeois geographical materialism (38). However, only orthodox Marxism can see the true complete "picture," the "inner order of the facts of nature," and their relevance to human society (53). Orthodox Marxism is a form of seeing that is holistic, penetrative, and transcendental.

The equating of sight with truth in Wittfogel enables his analysis to contain the problem of language by subsuming it under the problematic of seeing. The fact that Wittfogel's master categories of "man," "nature," "production," and "labor" are socially signified categories with ambiguous, incomplete, and unrealizable identities that vary historically is not acknowledged by Wittfogel's text. Such categories function visually, not linguistically, for Wittfogel. They are held to be unproblematically visible. They are immanently recognizable and transparent. Nature is always nature. It is simply a matter of seeing it completely and holistically.

Given the above arguments, we need to reevaluate the merits and claims of Wittfogel's attempt to develop a critical theory of geopolitics. There is much that could be said about the failure of Communist intellectual practices at this time, failures that were to prove fatal to the working-class movement as Germany moved from crisis to fascist rule. One of these failures was the growing philosophization of Marxism. The striking fact about Wittfogel's theory of geopolitics is that it was a theory of the theoretical status of geopolitics within Marxist thought rather than a theory of how geopolitics actually worked in Germany. Geopolitics was read in a crude economistic way, an epiphenomenon that was not of much interest in itself. Leading intellectuals, like Haushofer, were mere epigones of an old geographical materialism that served bourgeois interests. Wittfogel's analysis did not actually engage with geopolitics as a social phenomenon. The complex questions raised by Haushofer, the establishment of a school of German *Geopolitik,* its relationship with the Nazis, and the popularization of concepts such as lebensraum were not addressed by Wittfogel. The complexity of geopolitics was collapsed into categories that actually prevented an understanding of geopolitics. Reading geopolitics as bourgeois ideology attributed a coherence to geopolitics that it did not have. Haushofer's geopolitics was, if anything, closer to an aristocratic viewpoint than anything else.[10] Like other conservatives, Haushofer was actually quite hostile to "bourgeois" values and the "bourgeois" Weimar Republic. It was the coming together of a certain aristocratic weltanschauung with the antimodernist counterrevolutionary zeal of the Nazis that made German geopolitics significant ideologically. Wittfogel missed this entirely, in 1929 at least. Behind his local theoretical failure was the larger failure of Stalin and the Communist movement to appreciate the dangers of fascism until it was too late.[11]

A second observation that can be made about Wittfogel's analysis is that it is not an antigeopolitics. Wittfogel did not seek to break from the possibility of a geopolitics of human history but to offer a different and better form of geopolitics. Wittfogel shared many of the same epistemological assumptions of German *Geopolitik.* Both believed in the possibility of the objective study of human history. The goal of research was to move beyond mystifications to establish truth in all its clarity. Both held a "natural attitude" (in Husserl's sense) to human history, analyzing it from a position above and beyond it. In the conclusion to his article, Wittfogel quotes Georgii Plekhanov (to whom his article owes a great deal) approvingly to the effect that only an investigation that combines both the natural and the social conditions of production can disclose "the innermost secrets of history." This same will to uncover "innermost secrets" characterized German *Geopolitik,* and it has remained a persistent feature of intellectual practices that write under the sign "geopolitics" since then.

The fact that Wittfogel and Haushofer shared certain fundamental epistemological principles accounts for one of the more curious turns in the history of geopolitics, namely the republication of excerpts of Wittfogel's article in the *Zeitschrift für Geopolitik* in 1932.[12] Haushofer followed Wittfogel's works with great interest (given his own interest in Asia) and consistently gave them enthusiastic reviews in the *Zeitschrift.*[13] In republishing Wittfogel, Haushofer noted that although he was a Communist, his "basic position has scientific value."[14]

While Wittfogel and Haushofer may have shared certain epistemological principles, their practical politics were far apart. According to Ulmen, Wittfogel was the only member of the Frankfurt Institute to abandon his scholarly work in 1931 to fight the Nazis (146). In that year and the next, he produced a series of engaged articles on Hitler, Italian Fascism, anti-Semitism, and the "fascistization of all ideology" by the Nazis. Fascism was still read in economistic terms—its mysticism expressed the spiritual bankruptcy of the bourgeoisie—but Wittfogel demonstrates a growing awareness that the "decline of science" evidenced in fascist ideology was associated with a "decline of the bourgeois world." This made it possible for brutality to be a principle of social order (151–52).

In 1933 Karl Haushofer marked the Nazi seizure of power with the publication of the pamphlet "National Socialist Thought in the World." Haushofer was put on the radio, and his "World Political Survey" was broadcast every month by all radio stations throughout Germany. Wittfogel, by contrast, attempted to escape Germany and was arrested and moved through a series of different concentration camps. Wittfogel's wife tried to intercede with Haushofer on Wittfogel's behalf, only to be told that those who had "lost the game" now had to suffer the consequences. Nevertheless, Haushofer did put a word in on Wittfogel's behalf with Hess, to no immediate avail, although, in late 1933, Wittfogel was released from his internment as he lay sick with rheumatism brought on by forced labor in the concentration camps (162). Wittfogel and his wife left Germany a few weeks later. Although his critical approach to geopolitics had its flaws, Karl Wittfogel fought fascism in an intellectual, political, and personal way, a fight many others chose to avoid.

ISAIAH BOWMAN: GEOPOLITICS AS PSEUDO-SCIENCE

The political activism of Isaiah Bowman was of a very different sort from that of Karl Wittfogel. Trained in physiography by William Morris Davis at Harvard, Isaiah Bowman's intellectual interests and social position were transformed by U.S. involvement in World War I. Because of his position as director of the American Geographical Society, Woodrow Wilson appointed Bowman the chief territorial specialist of the U.S. delegation to the Versailles Peace Conference. Through the contacts he made at the conference, Bowman gained access to the elite social circles of the American establishment. Out of the conference grew an effort, by British and U.S. delegation members, to create an international research organization, with branches on both sides of the Atlantic, that would promote the ideals of internationalism and Anglo-American cooperation in public life in both states. The effort, however, did not develop in a coordinated way. The British did establish their branch of the proposed organization, calling it the Royal Institute of International Affairs, but the American branch floundered. Political circumstances were running strongly against Wilsonian internationalism in the United States as Congress rejected U.S. participation in the League of Nations.

One group that found the growing political isolationism of the U.S. state disturbing was the informal dinner club of East Coast lawyers, bankers, and academics that Elihu Root, secretary of state under Theodore Roosevelt, led since 1918: the Council on Foreign Relations. In 1921 Isaiah Bowman and those others associated with the floundering American branch of the Institute of International Affairs merged with Root's dinner club, keeping the name the Council on Foreign Relations.[15] The explicit aim of the council was to influence U.S. governmental and elite public opinion toward internationalism. An activist, internationally engaged U.S. state was not simply an ideological aspiration; it was also very much in the economic self-interest of members of the council.

To promote this end, the council founded the journal *Foreign Affairs* in 1922, with Isaiah Bowman actively involved as a member of the editorial advisory board. *Foreign Affairs* was designed to foster a new geographic consciousness of the United States and its role in foreign affairs among public opinion makers and influential governmental officials. In a private letter Bowman wrote that the journal was "a plea for a forward United States foreign policy, interested in exploiting the world's natural resources and putting affairs in Washington in the hands of dispassionate experts who, unlike the public at large, know what they are doing."[16]

Not surprisingly, Isaiah Bowman saw himself as one of these "dispassionate experts." In 1921 Bowman published *The New World: Problems in Political Geography*, a panoptic survey of the empires, states, and colonies of the world in the wake of the territorial rearrangements after World War I. The volume quickly went through four editions. In the preface to the fourth edition, Bowman explicitly states the political value of the work:

> To face the problems of the day, the men who compose the government of the United States need more than native common sense and the desire to deal fairly with others. They need, above all, to give scholarly consideration to the geographical and historical materials that go into the making of that web of fact, relationship, and tradition that we call foreign policy. As we have not a trained and permanent foreign-office staff, our administrative principles are still antiquated. Thus even the loftiest intentions are too often defeated. To elevate the standards of government there is required a continuous examination of contemporary problems by citizens outside of the government service. In this way new points of view are set up and independent judgements made available.[17]

The existent common sense of U.S. foreign policy officials is revealingly described as "native." To Bowman, this signified parochialism and isolationism, a sense devoid of worldly knowledge. The scholarly study of political geography can, however, provide career foreign policy officials with a desirable, modern, "global view." Interestingly, the governing metaphor is that of elevation. Native common sense may have lofty intentions, but too often they are defeated and it remains base and parochial. Geographical education is associated with an *elevation* of the standards of government and the development of new points of view, which are, on the evidence of *The New World*, panoptic: detached, overlooking, all-encompassing, and surveying. Geography, for Bowman, was about seeing the world as a unitary space. U.S. foreign policy officials needed training in this way of seeing.

Bowman was no doubt pleased that the U.S. Department of State distributed four hundred copies of *The New World* to U.S. consular offices throughout the world. Neil Smith notes that approximately eighteen thousand copies of the English-language edition were sold and that as late as World War II the U.S. army distributed two thousand copies to its camp libraries.[18]

Although Bowman conceived of *The New World* as serving an important political need, he nevertheless represented it as a scholarly work. In Germany, however, *The New World* was read with some bitterness as a geographical manifesto written by one of the architects of the postwar territorial order. The Anglo-American panoptic it brought to bear upon the problems of the new world order was explicitly challenged by those associated with the *Zeitschrift für Geopolitik*. In 1934 a working group of German geographers published a three-volume study, *Macht und Erde* (Power and the earth), which was described by one of the contributors (Otto Maull) as the German answer to *The New World*.

With German geographers seeking to emulate Bowman as Germany came under Nazi dictatorship, it was probably inevitable that, once war broke out and Allied propaganda had catapulted German *Geopolitik* to public prominence, the practice of geography by Bowman and the practice of geography by figures like Haushofer would be linked. "Geopolitics," as we saw in the previous chapter, became a convenient and fashionable word: all forms of work by geographers for the state could potentially be dubbed "geopolitics." In public discourse, many began to refer to Bowman as "our" geopolitician, including anti-Nazi thinkers whom Bowman counted within his own ranks.[19] This development greatly irritated Bowman, as did the emergent tendency to refer to him as "America's Haushofer." Some analyses even declared that American geopolitics was actually a precursor of German geopolitics (thus making Haushofer Germany's Bowman!). The erosion of the political legitimacy of Bowman's geographical practice by the term "geopolitics" demanded a response. This took the form of a polemical broadside "Geography vs. Geopolitics" published in the *Geographical Review*, the journal of the American Geographical Society, in October 1942.

In "Geography versus Geopolitics" Isaiah Bowman codified and starkly articulated a critical approach to geopolitics that was to be the governing understanding of "geopolitics" within Anglo-American geography until the early 1980s. Bowman's critique of geopolitics was neither novel nor intellectually sophisticated. Its argument was of a kind with the propagandistic type of intellectual practices found in the United States during the war. But the article was, first and foremost, a personal defense. Most of the article is taken up with substantial discussions of the merits of Bowman's past work and his previously published reviews of German geography, which are grouped under the heading "Forewarning Recalled."

The strategy of Bowman's defense is to equate his own reputation with that of geography, science, American democracy, and the rights of the individual. German geography blurs to become "geopolitics," which is then associated with pseudoscience, expansionist imperatives, Nazi dictatorship, and worship of the state:

> Geopolitics presents a distorted view of the historical, political and geographical relations of the world and its parts.... Its arguments as developed in Germany are only made up to suit the case for German aggression. It contains, therefore, a poisonous self-defeating principle: when international interests conflict or overlap *might* alone shall decide the issue.[20]

Geopolitics, in Bowman's text, is a signifier without an identifiable referent. It becomes an abstract Other, an outside, a strategy of the enemy. What is ironic is that it was this strategy that presented a "distorted view." At this very time, Haushofer's son Albrecht was in prison because of his association with Hess's flight, while his father was viewed with considerable suspicion. Haushofer's writings had been a sustained argument for an alliance with, not a war against, the Soviet Union. In Bowman's text, however, German geography, German theories of government (reduced to Treitschke), and Nazi ideology are treated as different instances of the same "crooked and evil philosophy."

The persuasive force of Bowman's analysis is dependent upon a wealth of conceits, the most prominent of which are the distinctions between geography and geopolitics, science and pseudoscience. Yet, maintaining a clear distinction between these concepts was not possible for Bowman. Bowman's "Forewarnings Recalled" discussion of German political geography records both his praise and his criticism of certain works. German geography (in this case the work of Alexander Supan) was capable of being both "excellent" and "illogical" at the same time. A writing could therefore be both a work of geography and a work of geopolitics at the same time. It could be both scientific and pseudoscientific, by Bowman's own logic.

Bowman defends *The New World* as being a nonideological, scientific work. "It interposed no ideological preconceived 'system' between a problem and its solution in a practical world in which historical accident, not design only, had played so large a part. It sought to analyze real situations rather than justify any one of several conflicting nationalistic policies" (653). Ideology, for Bowman, was a system that justified or rationalized certain foreign policies. Yet, Bowman's Wilsonian internationalism (which he had modified considerably by World War II) was not judged to be an ideology. In other words, Bowman makes the ideological claim that his ideology is not ideological. He uses the concept of "ideology" in an ideological way to normalize his own ideology as objective. When Bowman looked at the world, he saw it as it really was. The problem of language, of the social limitations of the categories used to constitute the world (for example, the categorization of things into "historical accidents" and things that are "design") is evaded. The surface of the new world was taken to be immanently meaningful to the reasonable, dispassionate gaze of a geographical expert like Bowman.

We can make a similar argument about his strategic use of the term "geopolitics." Bowman concludes his piece by stating that geopolitics is "simple and sure, but, as disclosed in German writings and policy, it is also illusion, mummery, an apology for theft" (658). Note the implication (already suggested earlier when Bowman used the phrase "as developed in Germany") that geopolitics can be "disclosed" in other ways from that found in Germany. Geopolitics, in other words, is a class, one member of which is *Geopolitik*. Bowman sounds a warning in his article about the migration of geopolitics from Germany to America (in a veiled reference to the contro-

versial maps of George Renner). But the most prominent example
of American geopolitics at the time was Nicholas Spykman's *America's Strategy in World Politics,* which was published the same year.
In an earlier edition of the *Geographical Review,* Bowman had himself reviewed this work and lavished it with extravagant praise. "On
grounds of merit and public value *America's Strategy in World Politics* should be read in not less than a million American homes. Every
government official responsible for policy should read it once a year
for the next twenty years."[21] In practice, Bowman was not an opponent of geopolitical thinking. Indeed his distancing himself from
the term "geopolitics" was part of his own geopolitical practice. By
dubbing the enemy's foreign policy as "geopolitics" he normalized his
own geopolitics as "scientific geography." Soviet geography adopted
a similar strategy after World War II.[22]

We can get some indication of the importance Bowman attached
to "Geography versus Geopolitics" from the fact that he distributed several hundred copies of the article not only to other academics but to business and political leaders as well.[23] At the time of
the publication of "Geography vs. Geopolitics" in October 1942,
Bowman had assumed greater governmental responsibilities and was
deeply involved in U.S. planning for the postwar world order, particularly the fate of the British Empire. In early 1940 Bowman had
declared that the answer to German territorial lebensraum is economic lebensraum for all. In practice, this meant a global lebensraum
for American business, and it was this goal that Bowman resolutely
worked to achieve throughout the war years.[24]

"Geography versus Geopolitics" is best understood as a work of
wartime propaganda. Bowman made war on Germany in the text.
It is significant in the history of critical approaches to geopolitics
for two reasons. First, it helped codify the dominant strategy by
which postwar Anglo-American geography (and probably geography in a great deal of other countries as well) handled the profound
and disturbing questions raised by the phenomenon of geopolitics.
Geopolitics was a category that demarcated that which geographers
did not do. It was the dangerous outside that defined the scientific
inside that was geography. It was the pseudoscience that not only
legitimated but made the real science of geography necessary. For
this strategy to work, geopolitics had to remain a composite abstraction, a practice without definition or specification, a word that

simply signified that which was illegitimate. By this eminently political move, the politics of geography and geographical knowledge
were elided. One consequence of this was that certain conservative
(and even profascist) German geographers and German geographical works were easily reassimilated into the community of Western
"science" after the war.[25] The best example is the work of Walter
Christaller (1893–1969) on central place theory; his work was enormously influential within the discipline after the war. Christaller's
1933 doctoral dissertation, "Central Places in Southern Germany,"
was considered one of the founding texts of modern theoretical (that
is, positivist and deductive) geography. It was not generally noted,
however, that Christaller—who joined the Nazi Party in 1940 and
then the Communist Party in 1945!—worked during the war on
developing and refining his work within the context of drawing up
blueprints for the Germanic settlement of conquered territories in
the east, principally Poland.[26] His work exemplified the "reactionary
modernism" of the Nazi state—its mix of modern technological
and bureaucratic planning with the antimodernist romantic ideals
of an idealized German past—for it sought to rationally plan idealized, organic village communities of German settlement on territory
seized from Poles and cleansed of Jews.[27]

Second, the consequences of this were that postwar geography
never actually confronted the questions raised by the use of geographical knowledge during the war. The role the intellectual discipline of geography played during the war was considerable on all
sides of the conflict. The role of popular geographical identities and
slogans was even more significant and deserved serious research.
Popular terms like "lebensraum," "Greater East Asian Co-Prosperity
Sphere," and "Atlantic community" naturalized highly contentious
geographical processes and communities.[28] In order to wage war,
geographical images of one's own place and that of the enemy had
to be starkly drawn. In Germany, Ufa newsreels and Reichsfilmkam-
mer documentaries like *Der Feldzug in Polen* (Campaign in Poland,
1939) and *Sieg im Westen* (Victory in the West, 1940) pioneered the
use of animated cartography (bleeding maps, thrusting arrows, en-
circling military movements, and so on) to foster the requisite public
consciousness about the war. In the United States, such techniques,
as we noted in the last chapter, were copied to convince geographically innocent soldiers to die for places they might never have heard

of. The world the 834th Photo Signal Detachment of the U.S. army projected in its *Why We Fight* series was a Manichaean one of slavery and freedom, gangsters and good guys—a world of dark and light hemispheres. The ability to control the production of geographical identities and knowledge during World War II was crucial to the conduct of the war.

None of these eminently geographical issues were ever confronted and problematized by a newly sensitized political geography. To problematize them would have meant asking dangerous and destabilizing questions, questions not simply about the (ab)use of geographical knowledge in Nazi Germany but about the relationship of geographers and geographical knowledge to political power in general. More profoundly, it could have led to the questioning of the status of science and the functioning of reason within a modernity that had facilitated a program of genocide against the Jews and a horrific, machine-driven mass slaughter inadequately described as "war."

Such issues and questions, however, were never confronted. Under the guidance of Richard Hartshorne, political geography in the postwar period retreated from anything that appeared political and controversial. Yet, Hartshorne was in a unique position to lead research into modernity and the politics of geographical knowledge. In the fall of 1938 Hartshorne had gone to Vienna and found himself in post-*Anschluss* Austria. Hartshorne lived in a Jewish apartment block, and his correspondence records the arrest of some Jewish neighbors. Nevertheless, he remained remarkably naive to the emergent political situation, noting how "it is wonderful to live in a country in which discipline is so splendidly developed."[29] Back in the United States in 1942, after the publication of his book *The Nature of Geography* in 1939, Hartshorne headed a Projects Committee in the Research and Analysis Branch of the Office of Strategic Services (OSS). In this bureaucratic setting, Hartshorne deepened the commitment he had expressed in *The Nature of Geography* to objectivism, and his stance within the OSS was explicitly perceived as representative of "objective science" (312).

After the war, Hartshorne ascended to a position of significant influence within the discipline of American geography. In 1950 he surveyed the field of political geography and found much that he considered to be explosive and dangerous:

We may have produced no atom bombs in political geography, but the field is nonetheless strewn with dynamite—it is no place for sophomores to play with matches. Fortunately, we appear to have escaped the danger of repeating, in American terms, the crime of those of our colleagues in Germany who were responsible for the dangerous doctrines of geopolitics. But we will be exposed to similar dangers until the foundations of our knowledge in this field are on a much firmer basis than appears now to be the case.[30]

The field was "unorganized" and the least scientific of the branches of geography. Its practitioners were often poorly trained and sometimes published "misinformation or irresponsible recommendations proporting to represent more than the personal views of the author" (103–4). In certain cases, political geography was criminal. Given this situation, Hartshorne set out to provide the field with certain organizing principles that would "establish knowledge on such firm foundations that argument disappears, and acceptance becomes relatively enduring." Political geography was to be given a firm foundation and a solid structure of knowledge that would enable it to "arrive at applications of sound value in the solution of actual problems" (104).

The vision of transforming a minefield into something solid, firm, and consensual was a reflection of reconstructionist times (not to mention a gendered understanding of science). It was a vision in keeping with the turn U.S. political culture had taken in the postwar period. The politics of the New Deal, a politics based on explicit class appeals, ended with the declaration of war and the exhortation to national unity that accompanied it. Postwar politics came to be organized around what Alan Wolfe terms "the politics of growth," an apolitical form of politics that promised a harmonious, prosperous future for all social classes.[31] Geography within the United States developed its own similar politics of growth: growth and scientific progress could be achieved by explicitly rejecting the overtly political and by an active containment of the disturbing questions raised by the war. Hartshorne's attempt to develop a functional approach in political geography in 1950 was an attempt to create a political geography without politics, a diligent, scientific, civic-minded knowledge that was uncontroversial and consensual.[32] Integral to this project was the forced marginalization of geopolitics, the errant practice that was now best avoided and forgotten.

Outside the discipline of geography, the reaction to geopolitics was quite different. While some theorists followed the "pseudo-science" line and condemned the word while working within the problematic it marked, others readily embraced both the word and the practices associated with it.[33] Some within political science attempted to objectivize geopolitics by translating it into "geopolitical hypotheses," which were then put through a ritual of being "tested."[34] A few lonely figures in political geography continued to address global issues, but in a careful and circumspect way that carefully avoided the accusation of "geopolitics."[35] Although none of this intellectual work was outside politics and ideology, all claimed the mantle of science and objectivity.

YVES LACOSTE: GEOPOLITICS AS A FORM OF GEOGRAPHICAL REASONING

Although a widely disseminated and culturally hegemonic set of geographical identities concerning "the East" (communist, totalitarian, and enslaved), "the West" (free, democratic, and individualistic), and the "Third World" (the zone of conflict between capitalism and communism) were fundamental to the functioning of the Cold War, the social production of such geographical scripts was never investigated by the discipline of geography. It was not until contradictions began to develop in the postwar politics of growth ideology, with the civil rights movement and the Vietnam War in the 1960s, that questions began to be raised anew about geography and its relationship to a social order that was violently suppressing domestic minority rights while spending billions of dollars conducting a brutal war against an undeveloped nation thousands of miles from the United States.

As we have already noted, the social and political upheavals brought on by the civil rights movement, the Vietnam War, and new cultural movements led to the development of a radical geography in the late 1960s in the United States. The journal *Antipode* was founded as a journal of radical geography at Clark University in Worcester, Massachusetts, in 1969. It provided a forum that challenged the objectivist, value-free, political neutrality pretensions of geography as a science, an epistemological ideology that had taken a quantitative and technocratic turn in the late 1960s. Despite the significance of the Vietnam War to the political conscientization of a new generation of geographers, the new radical geography of the 1970s had disappointingly little to say about geopolitics in a substantive, empirical way. A naive rediscovery and enthusiasm for the dogma of old Marxist theoretical debates on capitalism and imperialism precluded actual empirical investigations of the Vietnam War.[36] The Vietnam War was read by one radical geographer as a consequence of "late capitalism's dependence on imperialistic domination and exploitation for its continued functioning and the inevitable reaction (in the form of liberation struggles) against this domination in the Third World."[37] In most instances, analysis remained at this level.

One notable exception was the work of the French Communist geographer Yves Lacoste. Like many of the new radical geographers, Lacoste's ideas and work are a product of a confrontation with the problems of the violent process of decolonization at this time, problems that were particularly intense in France because of its violent war against Algerian independence and its humiliating defeat in Vietnam in 1956.[38] As a member of an International Commission of Inquiry into War Crimes, Lacoste visited North Vietnam in 1972 and undertook an empirical investigation of the systematic U.S. bombing of the dikes on the Red River in North Vietnam, a bombing that appeared designed to destroy the irrigation system of the Red River Delta and so flood the homes and crops of the ten million people who lived in the delta region.[39] Lacoste's research on the Red River dike bombing strategy of the U.S. war machine in Vietnam was a brilliant example of a countergeopolitics, a writing from below on the environmental consequences of U.S. geopolitical practice in Vietnam.

Unlike so many other forms of radical geography, Lacoste's work is characterized by a sustained consideration of the meaning of geography and geopolitics. What distinguishes Lacoste's work from traditional political geography is, first, its appreciation of geography as a language and form of power/knowledge. Geography in Lacoste's work is never naively taken to be immanently meaningful or obvious. Rather geography is a social discourse, "a mode of representing the world" (244). It includes not only teacher's geography and academic monographs but mass-media-generated "geographical cliches and images," what Lacoste called, influenced no doubt by the Situationists, the screens of the spectacle of geography.[40] Geography is not only taught in classrooms but is also projected at the citizen-spectator in films, newspapers, advertisements, postcards, and travel brochures. The range of what is given to be geographical varies, however, and is dependent on what the existent social order wants to demarcate as "geographical."

Geography, for Lacoste, is, second, a strategic form of knowledge, in the words of his 1976 study, *La Géographie, ça sert, d'abord, à faire la guerre.* Before being made, in the late nineteenth century, into a discipline to be taught in schools (a move that obscured it as a strategic form of knowledge and gave it an innocent and neutral veneer), geography was a form of knowledge taught to kings, princes, diplomats, and military leaders:

> Geography is first and foremost a strategic knowledge which is closely linked to a set of political and military practices; these practices demand that extremely different, at first sight heterogenous pieces of information should be brought together. You cannot understand the grounds for existence nor the importance of such information if you confine yourself to the validity of knowledge for knowledge's sake. These strategic practices make geography necessary, primarily for those who control the machinery of the state. Is this really a science? It does not really matter; the question in not fundamental insofar as one is aware that geography, being the structuring of knowledge relating to space, is a strategic knowledge, a power.[41]

In designating geography as a strategic form of knowledge, Lacoste sought to undermine the scientific image of geography within France, an image promoted by the actually quite political Vidalian tableaux tradition — we must remember Vidal's efforts to secure the territorial integrity of Alsace-Lorraine within France and not Germany — as "the science of places" or the "new geography" associated with the introduction of quantitative methods into the discipline in the 1960s. Rather, geography is a form of knowledge that is part of the apparatus of the state's control of space and territories. Geography has existed ever since the apparatus of the state has existed. It has aided not only in the moving of troops, the location of fortifications, or the organization of traffic routes but also in the general mastering of space. Lacoste terms Clausewitz's *On War* a genuine work of "active geography" and Vauban one of the best geographers of his time (16). Military science and geography are inseparable.

Yet while the polemical title of Lacoste's book foregrounds the close connections of geography to the military and the practice of warfare, his argument does not represent it as only a military form of knowledge. Geography is also a practical form of knowledge that is necessary for territorial administration and government. Lacoste argued that it is a knowledge born out of the practical management problems of government, problems addressing the administration, surveillance, and control of populations, territorities, and colonies. The specific types and methods of geographical knowledge should be understood as part of the functioning of relations of power. The establishment and teaching of the discipline of geography in secondary schools, for example, is part of an attempt to inculcate a national ideology into the population, an effort to reinforce the notion of the nation as an intangible given (46). The map, Lacoste points out, is an instrument of power, a way of representing space that facilitates its domination and control. To map is to "formally define space along the lines set within a peculiar epistemological experience; it actually transposes a little-known piece of concrete reality into an abstraction which serves the practical interests of the State machine" (244–45). The survey is another type of geographical knowledge linked to state power. Geographers, Lacoste argues, have historically been "information agents." Down through the centuries they have paid attention to phenomena that are of potential use to military and governmental leaders: "Topography, for example, was described in terms of strategic and tactical interest; the distribution of population was described in terms of the administrative and political organization of space. The central notion of region, so perennially used in the discipline, derives from the Latin word *regere* (to rule). Etymologically speaking, then, a region is by definition a military region."[42]

Lacoste's linkage of the "region" to the military was part of his attempt to dethrone the Vidalian tradition of geography within France, a tradition in which the "region" was a core organizational concept. Although it was more complex than it appeared, this tradition had effectively depoliticized geography, in that its regional focus precluded the appreciation of the differential spatiality that made places; its emphasis on place, not people, separated geography from other social sciences; and its stress on the concrete encouraged a contempt for abstract contemplation, theoretical reflection, and polemical debate. To overcome these considerable shortcomings, Lacoste and a number of other French geographers (based in the experimental university at Bois de Vincennes) established in March 1976 the radical journal *Hérodote,* edited by Lacoste.[43] The first issue (*Attention: Géographie!*) announced a plan to use geographical tools, methods, and practices to reappropriate geography for radical ends. Geography was not to be reformed but turned around against those in power who used it. It was to be taught in a new way; exposing the locational strategies of firms, "unmasking" territorial management, and establishing a typology of domination. This plan was conceptualized in explicitly military terms; it was a question of "epistemological warfare," of developing an alternative "combative geography" that would inform the practice of militants and trade unionists and be informed by it to better situate the enemy, to better know and to better choose the terrain of battle.[44]

The first issue of *Hérodote*'s war-of-position geography begins with an interview with a figure who was one inspiration behind it: Michel Foucault.[45] Foucault had spent two years in Vincennes after its establishment, but there is no evidence that he ever had more than a passing interest in the history of the discipline of geography. Indeed, from the evidence of the interview, Foucault was somewhat skeptical of the intentions of *Hérodote*'s editorial group. Nevertheless, by its conclusion, he seems to change his mind and makes the following remarks on power and geopolitics:

> The longer I continue, the more it seems to me that the formation of discourses and the genealogy of knowledge needs to be analysed, not in terms of types of consciousness, modes of perception and forms of ideology, but in terms of tactics and strategies of power. Tactics and strategies deployed through implementations, distributions, demarcations, control of territories and organizations of domains could well make up a sort of geopolitics where my preoccupations would link up with your methods.[46]

It is important to note here that Foucault is using "geopolitics" as a metaphor for the operation of power in general. The term "geopolitics" is inflated into a sign for an understanding of knowledge as a form of tactical and strategic warfare. Foucault's analysis suggests the possibility of developing a "geopolitics of geopolitics," which would be a genealogy of how the concept of "geopolitics" came to be formed and function within contemporary discourse, a notion that would be taken up later by Richard Ashley. However, Foucault acknowledges his ignorance of the precise surfaces of emergence and forms of this genealogy.

Lacoste, however, is well aware of the specificity of the genealogy of geopolitics, which he understands as a type of geographical reasoning. In orthodox use, this is a type of reasoning that appeals to geographical "evidence" and "imperatives" to justify the particular foreign policy of a state. Geopolitics is associated with the imperialistic foreign policy practice of a dictatorial ruler (a Napoleon or a Hitler) that attempts to impose a geopolitical plan presenting itself as corresponding to the "nature of things."[47] Geopolitics is also associated with the tradition of political argument represented by Mahan and Mackinder. The theses of Mahan and Mackinder, Lacoste argues, "rest more on historical evocations than on rigorous strategic thinking, based as they are on grandiose geographical metaphors of the Land and the Sea" (214). Such theses lack scientific value but have a significant "lyrical value" that enables them to misrepresent the conflict between the Soviet Union and the United States as a metaphysical conflict between land and sea.

Lacoste's stress on science and rigor point to a critical theory of geopolitics that relies on an unexamined notion of objectivity. Orthodox geopolitics is objectionable to Lacoste not only because it is associated with imperialism but also because it is a type of geographic reasoning that is hazy, erroneous, distorted, and simplifying. Although international relations cannot be studied in a perfectly objective way, Lacoste nevertheless believes it possible to "reduce the influence of certain ideological assumptions that lead us to think in Manichean terms. Above all, we must avoid thinking that problems are simple when they are very complicated" (215).

Despite his well-founded suspicion of scientific geography, Lacoste nevertheless champions a form of "true geographic reasoning" that he sometimes describes as "objective" and "scientific" (216). At different points in his work, Lacoste expresses what he understands to be involved in this commitment. First, there is the possibility of a pure seeing of international affairs. Lacoste holds strongly to the possibility of seeing things as they really are (215). The argument against Manichaeanism in international relations, he suggests, is a moral one, but "it also arises from a desire to be more effective, and to see the situation more clearly. For us, the geographer is perhaps, amongst the many observers of the world today, one of the best equipped to do so."[48] Retained within Lacoste is an unproblematized Cartesian perspectivalism, the dream of the geographer as a removed and privileged seer.

Second, there is the task of unmasking the fraudulent. Geopolitics in the mass media is frequently associated with "banal statements and outdated slogans." It is vital that geographers respond to the growing interest in geopolitics "in a more satisfactory and rigorous manner and that we unmask all fraudulent geopolitical stances" (3). Initially at least, Lacoste envisioned the critique of geopolitics as taking the form of a revelation of ideological simplicity in the face of complex reality. The depoliticized geography produced by the French university system has not found a mass audience. Geographers need to repoliticize their discipline and so take their place within the public arena as permanent critics of the mass media's tendency to reduce complex geopolitical issues to simple slogans.[49]

Third, there is the necessity to exhaustively document the complex spatial relationships that are to be found in international relations. Geographers must think of space as something that is "textured, extremely varied and very complex."[50] Only the exhaustive designation of the spatial configurations that shape international relations and a "sensitive analysis of the articulation between them will make geographical reasoning worthwhile and enable geographers to arm themselves better against the influence of ideological assumptions."[51]

Finally, Lacoste advocates a return to the work of the turn-of-the-century French anarchist geographer Élisée Reclus (1830–1905) for renewed inspiration. Reclus's work is an example of how to disengage geographical knowledge from its links with the state apparatus without, at the same time, eliminating the political. Reclus developed geography's effectiveness by enlarging the "geographical," by emphasizing the ambiguity of "progress," and by turning knowledge against the ruling classes. In doing so, Lacoste argues, Reclus "advanced geographical reasoning as a method of objective and scientific analysis of a broad spectrum of reality" (8).

Lacoste's commitment to a Reclusian-inspired enlargement of the "geographical" led him to make the case for a geopolitics beyond that associated with imperialism, an alternative anti-imperialist geopolitics that justifies independence, autonomy, and liberation.[52] This geopolitics is not a geopolitics of and for the state but a geopolitics of and for social movements. Later in the 1980s Lacoste added further dimensions to his "other geopolitics." In 1986 Lacoste and thirty-six other authors published a mammoth three-volume study on the evolution of the *région* in France. The study of the *région* was an example of "inner geopolitics," which was separate from yet also complemented the study of "outer geopolitics."[53] In 1988, in what is perhaps the clearest statement of Lacoste's commitment to a panoptic gaze, Lacoste outlines a distinction between a partisan and an aloof geopolitics. The latter form of geopolitics, the form Lacoste champions, is a geopolitics that "looks down" upon international issues. It is less directly committed, a detached perspectival form of geopolitics that observes conflicts and tries to understand the reasons for events.[54]

There are many dimensions of Lacoste's perspective (and perspectivalism) on geopolitics that I have not examined, particularly his empirical analyses of Islam, Euromissiles, and the sea.[55] Nevertheless, we have considered enough to gain a sense of the value yet also the difficulties associated with such a perspective. The first of the difficulties is that Lacoste has a Cartesian persecivalist understanding of reality. "Reality" is "out there," a complex but nevertheless graspable and capturable external world of objects independent of signification. The geographer is like the detached observer of a distant battle. With care and exhaustive documentation, he can see the world as it really is, can narrate the truth of things, can effectively represent the complex way things objectively are. The rationalist dream of a transparent world, unobscured by language, is found throughout Lacoste's texts.

Second, despite the significant advance of recognizing geography as, first, a social discourse and, second, a discourse tied to systems of power, Lacoste's work arguably ends up falling back into the very ideological system of epistemology he wished to challenge. Geography, for Lacoste, clearly can attain an objectivity and a scientificness. It can approach the dream of geopolitics, that of panoptic survey and divination.[56] Lacoste's description of geography as a *strategic* form of knowledge is thus not a radical problematization of the conditions of possibility of geographical knowledge but a polemical argument against its (ab)use by the military. Geography is too easily turned away from power in Lacoste (a matter of making maps for trade unions, not capitalist firms, for example, rather than deconstructing the map). In not challenging the possibility of objectively knowing and panoptically seeing the world, in not problematizing strategy as de Certeau does, Lacoste arguably leaves the epistemological infrastructure of geography/geopolitics intact. Assertions of rigor, exhaustiveness, complexity, and science do not necessarily break from the conditions of possibility of geography/geopolitics and guarantee an antigeopolitical practice. Despite suggestive observations concerning the complicity of geographical knowledge with relations of power, Lacoste's work ultimately fails to specify in a detailed way how geographical knowledge functions strategically as a form of power/knowledge.

RICHARD ASHLEY AND DISSIDENT INTERNATIONAL RELATIONS: GEOPOLITICS AS THE SPATIALIZATION OF GLOBAL POLITICS

The term "geopolitics," as I noted in the previous chapter, emerged as an object within U.S. political discourse during World War II and, as a consequence, inevitably was confronted and studied within the subdiscipline of international relations. International relations, as many commentators have pointed out, is an American social science. Stanley Hoffman has argued that the development of international relations (IR) as a field resulted from a convergence of three factors: intellectual predisposition, political circumstance, and institutional opportunities.[57] The intellectual predisposition Hoffman identified as peculiarly American was the deep-rooted belief in U.S. culture that all problems can be solved and that the way to solve them is by application of the scientific method. For Hoffman, the idea of progress through the deliberate application of reason to human concerns was a national ideology in the United States. The political circumstance he identified was the ability of the U.S. state after World War II to undertake a vast expansion in its educational institutions. Political science and its international relations scion were part of a system of social science knowledge production that sought universalist explanations and pragmatic solutions to age-old problems. The institutional opportunities were the result of the transplantation of a large community of intellectual immigrants from Europe whose personal experiences and philosophical predispositions, according to Hoffman, enabled them to jolt American social science out of its parochialness and educate it about power politics. Among these "wise and learned" émigrés were Hans Morgenthau, Arnold Wolfers, Klaus Knorr, Karl Deutsch, Ernst Haas, George Liska, and the young Henry Kissinger and Zbigniew Brzezinski. To these figures, many of whom saw their work, if not mission, as bringing geopolitics to North America, we can add Frederick Schuman, Nicholas Spykman, and Robert Strausz-Hupé, figures we have already encountered advocating an American version of geopolitics.

Beyond these culturalist and conjunctural explanations for the development of international relations was the much more significant structural situation of the United States as the preponderant economic and military superpower after the devastation of World War II. That the political realist readings of international relations gathered together in the writings of the émigrés Hoffman identifies became significant was a consequence of these scholars codifying what policymakers needed or, perhaps more important, were already thinking and practicing. Although the relations between political realism and political practice are not straightforward, the renditions of the nature of international politics offered by Wolfers, Morgenthau, and others codified and helped constitute—beyond the personal intentions of any of these intellectuals—the political imagination, hegemonic ambitions, and reasoning processes of the bureaucrats and politicians who dominated the U.S. state after World War II. As an eminently pliable and flexible system of thought, political realism was the commonsense regime of truth by which U.S. foreign policymakers, bureaucrats, and functionaries understood their activities and justified their actions on a day-to-day basis.

Yet it is worth questioning whether political realism was ever a singular, coherent, immanently meaningful approach to international relations beyond basic foundational banalities about power politics. As Rob Walker has suggested, political realism must be understood less as a coherent theoretical position than as a site of a great many contested claims and metaphysical disputes.[58] The proliferation of fractured, specialized, and highly technical realist regimes of truth since the mid-fifties within international relations—such as Sovietology, nuclear strategy, game theory, and, later, neorealism—generated a series of self-demarcated debates within the field between traditionalists and behavioralists, realists and neorealists, and, by the middle to late eighties, between so-called modernists and postmodernists.[59] Richard Ashley, a central figure in the last of these debates, made a notable theoretical intervention in the early and mid-eighties against the neorealism of such disparate figures as Kenneth Waltz, Robert Keohane, Stephen Krasner, Robert Gilpin, Robert Tucker, George Modelski, and others.[60] Interestingly, Ashley's sustained critique of neorealism in 1984, which draws heavily from E. P. Thompson's critique of Althusserian Marxism, is one that revalorizes the work of "classical realists" like Hans Morgenthau and Henry Kissinger because of their supposed respect for practice, history, contingency, and agency. Ashley's particular reading of geopolitics, which we will examine later, should be understood within the context of this relatively sympathetic, although not uncritical, appreciation of classical realism when compared with neorealism.

By the late eighties, Ashley was drawing heavily on French poststructuralist writings to develop an ambitious critique of the modernism of IR theory, thus creating space for new dissident IR theories and practices. Drawing upon Julia Kristeva's 1977 essay, "A New Type of Intellectual: The Dissident," Ashley outlines a project for international relations that eschews the language and subjects of universalism ("some universal 'we' whose proudly certain identity is in reality a projection of a diffuse fear of abjection") in favor of "the work of a dissident," the "work of thought."[61] Echoing Kristeva, Ashley writes that this is a work that becomes possible only when one cuts all ties and becomes a stranger to country, language, sex, to any notion of a sovereign identity of Man.[62]

Dissident IR, as Ashley and Walker describe it in their introduction to a special issue of *International Studies Quarterly,*[63] begins from a recognition of the "intrinsic ambiguity" of modern global life. Imposed upon the uncertainty, indeterminacy, undecidability, disorder, and turmoil of global politics are systems of meaning with sharply bounded identities, fixed categories, stable values, and commonsense meanings. Echoing the rhetoric of Gilles Deleuze and Félix Guattari, they identify "deterritorialized sites where people confront and must know how to resist a diversity of representational practices that would traverse them, claim their time, control their space and their bodies, impose limitations on what can be said and done, and decide their being." It is the struggle between sites of deterritorialization and efforts at reterritorialization that constitute the "plot" of global political life for Ashley and Walker. At deterritorialized sites, identity is never sure, community is always uncertain, meaning is always in doubt. Indeed, people here confront arbitrary cultural practices that work to discipline ambiguity and impose the effects of identity and meaning by erecting exclusionary boundaries that separate the natural and necessary domicile of certain being from the contingencies and chance events that the self must know as problems, difficulties, and dangers to be exteriorized and brought under control.[64]

The work of dissident IR is to problematize this "territorialization" of global political life, to question the boundary-producing practices of modern discourse that divide self from other, rationality from irrationality, the inside from the outside, and reasoning man from history.

Following Foucault's reading of Kant and the "critical limit attitude" of the regime of modernity, Ashley argues that modern discourse, in order to privilege reasoning man from history, must first constitute reasoning man as a being different from and set in opposition to history. This is enabled by Kant's distinction between what man necessarily is — a rational human being — and what he contingently does. Man thus acknowledges his necessary historical limitations while also asserting his ability to transcend contingent experiences. The former necessary limitations, those essential to man, "provide the transcendental conditions of man's reason, man's total knowledge, man's capacity to assert total mastery over himself and over the contingency and chance of his experience."[65] The figure of Man is at once the source and the limit of criticism.

Out of this general discussion of modernity, Ashley isolates the problem of inscribing a paradigm of sovereignty in global politics. Such a paradigm is the ground, the tacitly acknowledged historical limitation, that empowers sovereign subjects to be recognized as capable of reasoning, reflecting, and discoursing on global politics. It is that which provides the conditions of possibility for reason, the unproblematic, taken-for-granted point of departure for political practice. The paradox of the paradigm of sovereign man is that its production must assume that it already exists. In Ashley's terms, in the regime of modernity, "the performance of the most consequential of world political tasks must always start from the premise that it has already been successfully performed. A paradigm of sovereignty must be produced through practices that rely on an understanding that is already in place" (271).

Ashley reads statecraft as a practice of enframing sovereign man, a practice whereby a paradigm of sovereignty is asserted and bound-aried in the face of ambiguous, indeterminate, and unlimited tra-versal struggles. "Practices of modern statecraft work not primarily by solving problems and dangers in the name of a domestic popula-tion already given, but by inscribing problems and dangers that can be taken to be exterior to sovereign man and whose exteriority serves to enframe the 'domestic population' in which the state can be rec-ognized as a center and can secure its claims to legitimacy" (302). Modern statecraft, according to Ashley, is "modern mancraft," the "art of domesticating the meaning of man by constructing his prob-lems, his dangers, his fears" (303).

It is the operation of the paradigm of sovereignty in global poli-tics that dissident IR seeks to challenge and problematize. Follow-ing Derrida, Ashley locates the place of dissident IR on the border-lines or margins of modernity, the shifting paradoxical places (or "nonplaces") defined by modernity's articulation of centers of iden-tity and sovereignty. Borderlines or margins are deterritorialized sites, points where the sovereign presences of modernity reach a limit and begin to unravel. Poststructuralism does not ask where the bound-ary is or how the boundary is marked. Rather it asks, "How, by way of what practices, by appeal to what cultural resources, and in the face of what resistances is this boundary imposed and ritualized?" (311). Dissident IR explicitly repudiates any identification of itself as a new perspective or single philosophical approach to global pol-itics. It is not an alternative paradigm but a critical attitude.

It could be argued that dissident IR's ostensible repudiation of interest in defining a new orthodoxy is merely an effective rhetori-cal move on its part.[66] Ashley's insurgent storming of the capitols of IR theory is all the more stealthy for its ostensible disinterestedness in this very battle. Just as dissident IR teaches us to be skeptical of the discursive strategies of orthodox IR theorists, so should we be skeptical of the operation of the rhetorical system deployed by pro-ponents of dissident IR. While Ashley's work is undoubtedly a pow-erfully eloquent and sustained argument for poststructuralist ap-proaches in the study of international affairs, there are two features of his work that practically narrow the very openness dissident IR seeks to create and celebrate.

First, Ashley's interventions are confined to the realm of IR theory discourse. Ashley's enounced commitment to thought is a commitment to a relentless philosophical form of discourse. He himself underscores the point that poststructuralist discourse remains *theoretical* discourse (278). It is not interested in the meticulous examination of particular cases for purposes of understanding them in their own terms. The questions it asks are abstract and philosophically profound — pro*found*, first, in the sense that they concern the problem of foundations, and *pro*found, second, in that they concern that which works prior to all foundation.[67] The productions Ashley chooses to interrogate are the texts of academic IR theorists, figures like Hedley Bull, Kenneth Waltz, and Robert Keohane. The considerable energy Ashley devotes to dismantling the canon of these high priests of IR is undoubtedly important in unraveling the sovereignty principles that center global political systems. But, perhaps more relevant to the actual practice of global politics are the mundane and often chaotic workings of the ill-trained, half-trained, or narrowly trained officials that run the everyday practice of statecraft.[68] What was valuable about Ashley's 1984 critique of neorealism was its recognition, gleaned from Pierre Bourdieu, of the critical importance of practice as the art of competently managing the ambiguities and uncertainties of circumstances. Ashley applauds classical realism for its implicit competence model of political practice, a model characterized by a commitment to the necessary ambiguity of political reality.[69] His arguments anticipate a critical ethnography of statecraft developing out of Bourdieu's work. Yet Ashley's later work loses interest in the practical conduct of statecraft. For an insurgent theoretical practice committed to historicity and the importance of practice in international politics, Ashley's own form of dissidence is remarkably free from an engagement with the messiness of the practical conduct of statecraft. Dissidence takes the form of the critique of the classics of IR theory rather than the practical deconstruction of the discourses deployed in the practice of statecraft. What is ostensibly a radical challenge to the possibility of theorizing IR arguably works out to reinscribe the very genre it would problematize (although this is perhaps an unavoidable consequence of a necessary exercise). How dissident IR was given form in its initial articulations at least puts in doubt the possibility of cutting all ties and becoming a stranger to language (the language of IR theory or political theory in general), country (the American desire to produce "theories of international relations"), and sex (the IR master).

Second, the narratives and discursive strategies of dissident IR theory run the risk of becoming merely another form of discursive power politics within the IR community, another insurgency that settled down to become an orthodoxy. Dissident IR encourages the reading of global politics in terms of certain standard narratives that could become a new dissident lore in IR, complete with its "collectively recalled creation myths, its ritualized understandings of the titanic struggles fought and challenges still to be overcome in establishing and maintaining its paramountcy" (230). Associated with the application of dissident IR is a certain degree of intellectual elitism. In reading global political life in terms of the operation of the paradigm of sovereign man, dissident IR tends to mark its own exceptionalism by locating itself at the margins of this paradigm. This encourages the conceit that dissident IR is different and that other critical forms of politics are hopelessly mired in the paradigm of sovereign man. Rendering wide varieties of political protest against environmental destruction, political repression, or human rights violations as reprehensible because they evoke spiritual, Marxist, or humanist collective subjects, however, is a strikingly monological reading of the politics of social movements. Not only is such a reading strategy a form of intellectual elitism; it also fosters a comfortable sense of exceptionalism on the part of those who proclaim their work as not operating from the standpoint of some sovereign subject or center.[70] Dissident IR too easily presumed that one can "speak from the margins" and maintain a distance from paradigms of sovereignty. It remains caught in the rituals and critical games — rigorous analysis, unceaseless questioning, and so forth — of academia and, as a consequence, blunts its radical political edge.

To its credit, however, dissident IR anticipates and responds to these criticisms. Ashley and Walker identify a first phase of dissident scholarship that, first, did not radically break with the discipline of IR and, second, had "more than a hint of hubris."[71] Such scholarship was ironically heroic, conceived around the conceit that dissident IR "knows something or has access to a vocabulary or method that has hitherto been beyond the discipline's reach" (396). In its early articulations, dissident IR lent itself to interpretation as the new credo, the new deciphering system that holds out the promise not only of emancipation from old orthodoxies but of mastery over the domain of "international politics." Subsequent articulations of dissident IR have comprehensively put these early criticisms to rest with a series of excellent empirical studies and engagements not with the "grand theory" of IR or with "classic" political theory but with the everyday textuality of foreign policy, security studies, and political life in the modern world.[72]

Both the strengths and weaknesses of mid-to-late-eighties dissident IR are evident in Ashley's critical reading of geopolitics.[73] This reading is produced within the context of his attempt to outline the features of a Foucauldian genealogical perspective on international politics. Such a perspective, according to Ashley, does not pretend to "an 'apocalyptic objectivity,' a totalizing standpoint outside of time and capable of enclosing all history within a singular narrative, a law of development, or a vision of progress toward a certain end of humankind. Eschewing any claim to secure grounds, the appropriate posture would aspire instead to an overview of international history in the making, a view from afar, from up high" (408). The distinction Ashley draws here is an unstable one, one that pits a totalizing standpoint against an alternative that reads very much like a totalizing standpoint, a point that is removed, elevated, and detached. The problem here is Ashley's failure to problematize the rhetoric of vision and the ocularcentrism of Western epistemology. Indeed, the vision of vision is crucial to how he describes the genealogical attitude. "From a distant genealogical standpoint, what catches the eye is motion, discontinuities, clashes, and the ceaseless play of plural forces.... Seen from afar, there is only interpretation, and interpretation itself is comprehended as a practice of domination occurring on the surface of history" (408). After outlining the various features of a genealogical standpoint, Ashley notes:

> Like geopolitics, a genealogical attitude is preoccupied with motion, space, strategy, and power. Like geopolitics, also, a genealogical attitude is distrustful of all approaches that would accord to moral claims, traditional institutions, or deep interpretations the status of a fixed and homogeneous essence, a final truth, an underlying law, a relentless continuity, or an ultimate origin of international political life. (411)

Ashley seeks to strengthen his analogy by describing the genealogical "point of view" in explicitly spatial terms. The genealogical analyst approaches international politics as (1) an object, a site, and a product of ceaseless struggles; (2) a place where power and domination are secreted in normalized forms; (3) a field of practice upon which specific subjects emerge, secure recognition, and act; and (4) a set of strategies by which regions of silence are imposed, boundaries of practice are secured, subjects are legitimated, and domination is violently projected in the world. Given the apparent similarities between a genealogical and a geopolitical point of view, therefore, one can describe the application of a critical genealogy to international politics as an exercise in charting "the geopolitics of geopolitical space."

There are many different ways to criticize this maneuver by Ashley, but we will confine ourselves to two problems. The first is that Ashley's analysis does not actually engage with the con-textuality of geopolitics at all. "Geopolitics," in Ashley's text, is a free-floating signifier that is loosely associated with realist views of international politics, although this link is unspecified and undocumented. This disengaged, abstracted geopolitics is far removed from the historical genealogy we have examined up to this point. One has, thus, the irony of the argument of geopolitics-as-genealogy being premised on a forgetting of genealogy. This forgetting, it can be further argued, leads Ashley to make an egregious comparison between a Foucauldian genealogical attitude and geopolitics. Geopolitics, as already argued, is precisely about moral claims and deep interpretations that postulate a fixed and homogeneous essence, an underlying law, a relentless continuity to international politics. It is the very opposite of what Ashley claims. We can attribute this specific error to the larger problem we already noted of early dissident IR's failure to move beyond doing battle with the discursive strategies of certain academic IR realists and neorealists.

Second, Ashley's commitment to reading genealogy-as-geopolitics leads him to adopt the very discursive positionality he seeks to problematize and undermine. Ashley's very descriptions of the genealogical posture fail to problematize the status of seeing. In seeking to repudiate an "apocalyptic objectivity," he nevertheless recreates the very ground from which just such a practice works. The very positionality that would allow one to capture an overview, view from afar, see from up high is not sufficiently questioned by Ashley. Indeed, one can argue that the continued maintenance of just such a geopolitical positionality is precisely what early dissident IR produces. The hubris of traditional geopolitics is not repudiated but reproduced.

Confining our remarks only to these problems in Ashley's reading of geopolitics, however, is to miss the innovative importance and broader consequence of Ashley's argument. Building on Foucault's suggestions, Ashley's argument is a radical subversion of the very notion of geopolitics. In a manner similar to his subversion of the anarchy *problématique* of IR realists to open up international politics to the genuine question of anarchy (understood as the absence of any central ordering presence competent and empowered to represent and fix the boundaries of the state and domestic society),[74] Ashley uses geopolitics to radically pose questions of geo-politics, questions concerning the politics of spatial demarcation, administration, enframing, and domination. Geopolitics becomes the means by which dissident IR can pose questions concerning the spatialization of global politics. Critical analysis, Ashley concludes, must take up the never completed story of the "geopolitical domestication of global political space," "the story of the power political making, maintenance, administration, and transformation of the practical boundaries of sovereign state rule — the boundaries separating a rational domestic order from the recalcitrant world beyond its sway."[75]

Radicalized in this sense, geo-politics is not incidental to the critical investigation of international politics. Indeed, the very problematique dissident IR poses is a spatial qua geo-political one. In his elaboration of this problematique, Rob Walker identifies the production of the categories of "inside" and "outside" as geopolitical. At its most fundamental, geopolitics, for Walker, is the bounded geometric spaces of the here and there,[76] subject-object, self-world, inside-outside spatial imagination of the seventeenth- and eighteenth-century ontological traditions (125–40). Walker particularly stresses the coextensive operation of Euclidean geometry and Newtonian assumptions about absolute space in fashioning the way political life is framed. Linking the one-point perspective system of late-fifteenth-century Italian painting with the cartographic revolution of the sixteenth century, Walker suggests that the conventional Western political map is a consequence of a way of seeing that treats space as empty, homogeneous, and capable of being divided into clear, linear, inscribed parcels. Each of the separate containers of space specified is "imbued with a sense of independent integrity and internal homogeneity" (130).

Walker's work, like that of Ashley, is highly abstract and remains bound to a project of theorizing international politics even as it seeks to break from the way such a project has been conceptualized in the past. Like Ashley, he does not address the particular significance of the geopolitical tradition except in passing.[77] Together, however, both Ashley and Walker provide intellectual support for the type of focused genealogical analysis attempted in this book. While neither writer foregrounds the question of seeing and space, both nevertheless provide a valuable sustained deconstruction of the geopolitics of the textual strategies by which "international politics" is produced.

SIMON DALBY: GEOPOLITICS AS SPATIAL EXCLUSION

While Ashley's and Walker's critical reading of geopolitics remains abstract and caught within the problematic of theorizing IR, the critical investigation of geopolitics by the political geographer Simon Dalby is grounded in the engaged political writings of the late 1970s political pressure group, the Committee on the Present Danger (CPD).[78] Originally established in the late 1940s, this right-wing political lobby organization was resuscitated in 1976 by a group of neoconservative political analysts just after the election of Jimmy Carter as president of the United States. Drawing its ranks from the Ford administration's Team B intelligence review team, which believed that the CIA had consistently underestimated Soviet military strength (the exact opposite, as the CIA has now admitted, was actually the case, thus making Team B's estimates doubly exaggerated), the organization was a forum for paranoid and out-of-favor "hardheaded" security intellectuals like Max Kampleman, Eugene Rostow, Robert Strausz-Hupé, Richard Pipes, Colin Gray, and Paul Nitze. The writings of the CPD were influential not only in providing discursive ammunition for the New Right's persistent attack on the foreign policy of President Carter but in helping constitute the Manichaean foreign policy views of President Reagan. Many members of the CPD later found important national security jobs in the Reagan administration.

It is in the texts of the CPD's intellectuals that Dalby finds "a framework whose central organizing theme is geopolitical" (60). Before examining these texts, however, Dalby outlines his conception of geopolitics. Taking his cue from Ashley, he describes his work as "concerned with 'the geopolitics of geopolitical space'; how a particular series of 'security discourses' establishes an ideological space from which to dominate, exclude and delegitimate other discourses, and how this particular formulation of the world constructed the USSR as a threatening 'Other' " (16). Geopolitics, for Dalby, is a term that has "many meanings, often merging one into the other, but all have in common a general concern with the interrelationship of space and power" (33).

Using dissident IR, Dalby makes a series of connections that radically expand how space and power are normally conceptualized in orthodox IR and political geography. First, Dalby makes an epistemological argument about discourse and otherness. Otherness, he claims, is inherent in the analysis of discourse. "It involves the social construction of some other person, group, culture, race, nationality or political system as different from 'our' person, group, etc. Specifying difference is a linguistic, epistemological and crucially a political act; it constructs a space for the other distanced and inferior from the vantage point of the person specifying the difference" (7). Following Foucault's and Ashley's suggestive example, Dalby reads the very specification of difference and the production of discourse as geopolitical. Otherness involves exclusion, and exclusion, in Dalby, is inherently spatial.

Dalby then narrows his focus to consider the discourse of "security" in Cold War global politics. The rapid growth of state functions in capitalist states since World War II, Dalby suggests, has "expanded the need for ideological justifications of the functions of capitalist states." Of particular concern is the perpetuation of a permanent state of alert within the Western system that justifies the need for a formidable and expensive nuclear-armed "security state." The military preparations of such states "require the creation of a permanent adversary, an Other whose threatening presence requires perpetual vigilance. The highest political objectives of the state are now phrased in terms of the maintenance of 'national security,' a security usually defined in negative terms as the exclusion of the depredations of external 'Others.' "

Combining both the general argument about the production of otherness through exclusion and the more specific theme of state security leads Dalby to proclaim an "essential geopolitical moment," a "basic process of geopolitics":

> The exclusion of the other and the inclusion, incorporation and administration of the Same is the essential geopolitical moment. The two processes are complementary; the Other is excluded as the reverse side of the process of incorporation of the Same. Expressed in terms of space and power, this is the basic process of geopolitics in which territory is divided, contested and ruled. The ideological dimension is clearly present in how this is justified and explained and understood by the populations concerned; the "Other" is seen as different if not an enemy. "We" are "the same" in that we are all citizens of the same nation, speak a similar language, share a culture. This theme repeatedly recurs in political discourse where others are portrayed as different and as threats; it is geopolitical discourse. (22)

Dalby here seems to be suggesting that the very process of the production of identity and difference, of selves and others, is geopolitical. Speaking of a separate and discrete "geopolitical discourse," therefore, is not appropriate, since all political discourse is geopolitical discourse.

Yet, the operational understanding of geopolitics that Dalby uses in his textual analysis of the writings of CPD intellectuals shifts between a series of different readings of the notion of geopolitics. Added to the reading of geopolitics as the production of otherness and the reading of geopolitics qua security discourse as spatial exclusion is a third reading of geopolitics as "classic geopolitics," which is alternatively a determinist Ratzelian interpretation of the Soviet Union or a specific Mackinderian-Spykmanesque type of security discourse. Central to the plot of Dalby's book is the theme that the security discourses produced by the CPD — discourses Dalby analyzes as Sovietology, power politics, geopolitics, and nuclear strategy — all represent the Soviet Union as inherently a geographically expansionist power. "These security discourses produce a USSR whose actions are at least partly determined by its geographical location and the major influence this has had on its history; here is the theme of classical geopolitics." This, Dalby argues, is Ratzelian classic geopolitics (76). A different geopolitics, which also earns the identity "classic geopolitics" in Dalby, is the geopolitics of Mackinder and Spyk-man, which is understood in the text as a set of geopolitical argu-ments and an explicit language or lexicon of terms such as "Heart-land," "Rimland," and the like (104–21, 166).

We can, therefore, differentiate at least four different readings of geopolitics in Dalby's text. His practical treatment of geopolitics as a multilayered phenomenon accounts for, among other things, the fact that he claims, on the one hand, that the "central organizing theme" of CPD discourse is geopolitical, yet, on the other hand, that geopolitics is merely one of the four discourses he ascribes to the CPD (60). Geopolitics is thus both particular and general for Dalby; it is both part and whole, implicit and explicit, a specific set of security jargon and a general epistemological condition. As a con-sequence, geopolitics is an overloaded sign in Dalby's study, a sign that marks the general problematic of geo-power, yet a sign that also marks the particularistic question of the geopolitical tradition. The result is sometimes confusing, as when, for example, he writes:

> This [CPD] discourse of the Other is geopolitical in the sense that it creates an external (threatening) antagonist in a particular way *vis a vis* domestic political concerns. It is also geopolitical in that it is a particular exercise in geopolitical "geo-graphing" which draws on the traditional texts of Mackinder and Spykman to explicate a par-ticular geography of the Other, a geography which is interpreted in deterministic terms. This discourse of the Other is also geopolitical in the sense that it accepts the reification of political power in the particular relation of power and space of territorially defined states. In addition, this analysis shows how the overall logic of the discourse of the Other as constructed through texts on nuclear strategy, inter-national relations and Soviet history is structured on the classic geopo-litical conceptions of Spykman and Mackinder, a crucial point missing from other analyses of the CPD and of the Reagan administration foreign policy. (41)

There are, I would argue, two crucial difficulties with Dalby's treat-ment and analysis of geopolitics. The first is that, although Dalby's work is one of the most sophisticated critical treatments of geopoli-tics within political geography, it does not, ironically, have a devel-oped understanding of geopolitics as a sign. Dalby never sorts out the competing and crosscutting conceptions of geopolitics he entertains in his treatment of the CPD. Nowhere does he systematically sort out the distinction between epistemological geopolitics and security geopolitics, implicit geopolitics and explicit geopolitical language, the geopolitics of the production of Otherness and the geopolitics of Heartlands and Rimlands. One reason why there is an absence of a sustained reflection on the term in Dalby's work is that it is given over more to literature review than to thinking through the problem of how geopolitics is used. A subsequent article also prefers to review the literature of dissident IR than to explore the ambiguities of geopolitics as an object for critical theory.[79]

Nevertheless, Dalby's extension of some of the themes opened up to analysis by dissident IR is extremely suggestive and useful. The polyphonic reading of geopolitics found in Dalby empowers subversive readings of strategy, space, and power in ways that actively deconstruct the conditions of possibility of traditional geopolitical theorizing and practice. By multiplying the possible meanings of geopolitics, Dalby calls the very sign and the problematic of geo-power into question. Like dissident IR, Dalby uses the poststructuralist preoccupation with identity politics to foreground the problem of how difference is constructed. In a manner similar to the way Ashley and David Campbell redefine foreign policy as a boundary-producing practice, Dalby redefines geopolitics as a boundary-producing practice whose "essential moment" is the exclusion of the Other and the inclusion of the Same.[80] This essentializing of geopolitics within the terms of an identity politics narrative, however, is precisely the type of reductionist reading strategy Dalby is seeking to problematize, a way of reading that effaces the particularity of geopolitics. There is a danger, already evident in its very conceptualization, that the identity politics narrative becomes an abstract and reified formulaic reading strategy that allows one to read off the production of Otherness from certain foreign policy texts. The texts of the CPD become another example of "the Western structure of thought," "Western modernity" or "Orientalism."[81] One set of essential foreign policy abstractions is merely replaced by another.

Furthermore, to read geopolitics principally in terms of spatial exclusion, as the production of a territorially specified threat (160), neglects other possible and highly relevant readings of geopolitical discourse. Geopolitics, as I have already suggested, is a point of intervention into a multiplicity of different con-textualities. In producing Otherness, geopoliticians are not only specifying a dangerous external Other beyond the territorial borders of the state, as Dalby seems to imply. They are also projecting an image of their own subjectivity and its Other (the sentimental as opposed to the hardheaded theorist). The Other is not only a beyond but a within, a threat not only from abroad but also within the domestic and personal sphere. Reading geopolitics-as-spatial-exclusion solely in territorial terms limits other possible productive readings of this suggestion.

A second overarching difficulty with Dalby's reading of geopolitics concerns his method of the textual analysis of CPD writings. Dalby's approach is narrowly textual, preferring to read the texts themselves, in the manner of New Criticism, rather than the larger intertextuality—U.S. political discourse in general, the institutional position of CPD intellectuals, the personal biographies and past political involvements of CPD intellectuals, and so on—that gave them meaning. This decision, which Dalby defends at the outset, unfortunately impoverishes his readings of the texts, limiting the potential ironies, contradictions, and psychoanalytic observations Dalby could have made about the texts. More significant, however, is the way in which Dalby narrowly reads the texts. At the outset, in a typical dissident IR rhetorical flourish, Dalby declares that his concern is the postmodern concern "to leave power nowhere to hide" (5). Postmodernism is concerned with opening up possibilities and investigating points of struggle rather than charting the unfolding of a univocal History. Postmodernism is, presumably, against the monological reading of texts. Yet Dalby's strategy of reading is one that reproduces certain textual strategies that could be considered "modernist." Dalby represents his task as "a detailed unpacking of the logic of the security discourses used by the CPD." The CPD's nuclear strategy writings, for example, "are clearly decipherable only if their geopolitical premises are understood" (63). This notion of a critical reading as a clarifying and a deciphering is particularly important in understanding Dalby's strategy of reading. Three motifs are recurrent in his reading of CPD. The first is the persistent reference to "ideological moves," which are variously held to perform certain crucial legitimation, enframing, and exclusion functions in CPD discourse (48, 53, 63, 66, 69, 73, 84, 119, 159). The second is a recurrent understanding of the meaning of CPD discourse and its relationship to geopolitics in terms of essential "keys" (76, 82, 84, 100, 133, 158). The geopolitical premises of the CPD foreign policy, for example, are so crucial because they provide "the ideological keystone to their whole position" (119). A "geopolitical interpretation provides the key to Gray's theory of victory, a key that most commentators have missed" (157). The "particular specification of Otherness in terms of a geopolitical expansionist threat is the key element which articulates all the security discourses together" (158).

The operation of the first of these motifs assumes a rather questionable notion of power. Power is something that operates by being concealed. The task of the critical theorist, therefore, is to expose, reveal, and demystify it, to make it explicit (168, 180). Dalby wants us to see power; his goal is to make power visible. The operation of the second of these motifs treats geopolitics as the enigma that reveals all, the missing jigsaw piece that finally reveals the pattern (41, 148). For Dalby, the practice of critical geopolitics is the practice of the decipherment of enigmatic signs, the keying of locks. Once we complete this, then we can see. Dalby's attraction to such a narrative can be explained by his disciplinary situation, by his predisposition to find geopolitics as important, pervasive, and crucially explanatory, the concept that sets the terms of political debate (161). However, the combined effect of these two motifs does raise the question as to whether Dalby's reading of geopolitics actually reproduces some of the very strategies he aspires to critically analyze. The notion of deciphering a sign system to reveal the real nature of power to the eyes is already familiar to us from the geopolitical tradition, as is the tendency to attribute overarching explanatory significance to a latent factor (the "geopolitical") that is hidden or missing and needs to be made visible and present. Dalby's critical geopolitics, in other words, is partly complicitous with the very reading and writing strategies he wishes to problematize. He strives to produce a total reading of geopolitics, to uncover its keystone significance, decipher its underlying logic, expose its ideological moves, and render its power relations transparent to all.

Yet Dalby's text is much more than this. As the first systematic attempt to confront the textuality of geopolitics, its signifying excess is its success. Dalby treats the question of geopolitics as a question of signification and textuality. Although he does not systematically reflect on the metaphorics of vision that makes geopolitical discourse possible, Dalby is well aware that any aspiring critical geopolitics cannot replicate the Cartesian perspectivalism of modernist thought. Critical geopolitics, he suggests, "rejects the politics of grand detachment, the illusion of the Archimedean point from which the whole world can be grasped, in favor of critical disputations of the designations of reality specified by hegemonic discourses" (180). Critical geopolitics, in other words, puts the possibility and the politics of perspectivalism into question.

BEYOND THE PROBLEMATIC MARKED BY "GEOPOLITICS"

In this chapter, I have sought to problematize different strategies of reading geopolitics critically, some of which are only superficially critical, while others ask more profoundly challenging questions of "geopolitics" as the name for various historical congealments of geographical and political meaning. Disparate as they may be, the various reading and writing strategies reviewed in this chapter have the sign "geopolitics" as their point of departure. But, as should be obvious by now, the sign "geopolitics" does not have any essential meaning over and above the historical web of con-textualities within which it is evoked and knowingly used. As I noted earlier, we should not be mesmerized by the deployment and use of the sign "geopolitics" but look toward the more important problematic it marks. That problematic is the problematic of geo-politics, the politics of the production of global political space by dominant intellectuals, institutions, and practitioners of statecraft in practices that constitute "global politics." How is global political space envisioned and scripted by these actors? How are certain constellations of geopolitical meaning congealed around global visions like lebensraum, the Cold War, the New World Order, or global anarchy? How do certain locations within these global visions — Eastern Europe, the Middle East, Vietnam, Cuba, the Persian Gulf, Central America, or Bosnia — become the sights, sites, and cites of a governmentalized global scene? How, in sum, is geographical discourse governmentalized in the practices of statecraft by centers of authority and power? How is the spinning globe disciplined by a fixed "imperial" perspective, by mapping projects that reduce the indeterminacy of place to a homogenized surface of space? It is toward exploring these questions—the discursive and technologically mediated forms of geopower in the practices of contemporary statecraft—in a detailed manner, using the case of American foreign policy, that I turn in the final two chapters.

6

Between a Holocaust and a Quagmire
"Bosnia" in the U.S. Geo-Political Imagination, 1991–1994

> When I wake up every morning and look at the headlines and the stories and the images on television of these conflicts, I want to work to end every conflict.... But neither we nor the international community have the resources nor the mandate to do so. So we have to make distinctions.
>
> — ANTHONY LAKE, U.S. NATIONAL SECURITY ADVISER[1]

On May 3, 1994, President Bill Clinton took part in an unprecedented global discussion on world affairs from a television studio in Atlanta, Georgia. The event was organized by Cable Network News (CNN) and was broadcast to an audience of millions in more than two hundred countries and territories throughout the world. The president took questions from the local studio audience, which comprised over 160 international journalists from eighty countries, and from journalists in four remote television locations (Sarajevo, Jerusalem, Johannesburg, and Seoul) linked by satellite to the global forum. The spectacle was a global version of the "electronic town meetings" the Clinton campaign had pioneered during the presidential election of 1992. It was indicative of the new conditions of space and time shaping the conduct of American foreign policy, conditions where particular events in remote corners of the world are experienced immediately and instantaneously across the globe in real time. In shrinking distance and transcending national frontiers in an unparalleled way, global television networks like CNN promised a new transparency and visibility in world affairs. New visual technologies, from global spy satellites to mobile hand-held video recorders, enabled unprecedented visual documentations of natural disasters, war-induced famines, human rights abuses, and the machinations of military war machines. Never before did events across the earth's surface appear so close. With its planetary scope and worldwide audience, global television anticipated a global village *polis,* a global community of "public opinion" that would always be watching and judging the use and abuse of power in the post–Cold War world.

In noting how the information age and the end of the Cold War had transformed global affairs, President Clinton sought to outline a "clear road map" (what I would term a governmentalized geography) by which America would navigate its way through the new world disorder.[2] Responding to what he described as an "epidemic of humanitarian catastrophes" in places like Bosnia and Rwanda, he argued that "America cannot solve every problem and must not become the world's policeman." America has an obligation to join with others to relieve suffering, but America must, first and foremost, be guided by its own national interests and domestic priorities. Clinton's attempt to outline and explain his new geo-political road map for the United States did not, however, escape criticism during the global forum. In the most heated exchange of the evening, Christiane Amanpour, CNN's correspondent in Sarajevo, argued that Bosnia was much more than a humanitarian catastrophe; it was a fundamental question of international law and order. U.S. national interests, as the president himself conceded, were clearly at stake in Bosnia, yet Clinton administration foreign policy toward Bosnia is characterized, she suggested, by "constant flip-flops" that place the worldwide credibility of the United States in question. The president responded by noting that although his advisers told him to stay away from Bosnia because it was a "sinkhole," he felt the United States should be involved in the search for peace there and that his policy was not characterized by "flip-flops."

CNN's global forum and the heated exchange over Bosnia are a window into dilemmas and problems faced by U.S. foreign policy makers, given both the end of the Cold War and the rising intensity of informationalism in global affairs. During the Cold War, the U.S. governmental map of its strategic commitments, geopolitical interests, and moral obligations appeared relatively clear. Beyond a concern for its own territorial integrity and security in its Central American and Caribbean "backyard," the United States was committed by alliances to Western Europe and various states in Asia, the most important of which were Japan and South Korea. All these places were part of a governmentalized realm of a hegemonic U.S. state and, as a consequence, enjoyed *geopolitical proximity* to the United States. The United States was responsible for and had a primordial obligation to the governments in these states.[3] This governmentalized geography and geopolitical proximity was given military form by the stationing of U.S. soldiers abroad in strategic locations as a means of guaranteeing U.S. commitment to a certain geopolitical world order. Beyond a first order of Western European and select Asian places, the U.S. geopolitical engagement with the rest of the world decreased, although, during certain intense periods of Cold War competition, even relatively minor and distant geopolitical locations (like Vietnam, Angola, the Horn of Africa, and Afghanistan) took on strategic significance. After the experience of the U.S. state in Vietnam, however, the danger of relatively unimportant strategic locations taking on a significance greater than their intrinsic worth was recognized by the U.S. foreign policy establishment. Within the Pentagon in particular, such places were read through the Vietnam experience and coded as potential "quagmires," dangerous places that resisted the governmentality of the hegemonic U.S. state.

Since the end of the Cold War, the strategic geography of proximity and distance in U.S. geopolitical thinking has become diffuse and uncertain. With the disappearance of the "Soviet threat," the geo-political reasoning justifying the permanent stationing of U.S. soldiers abroad in locations like Japan and Western Europe (although not South Korea) has faced attack from a variety of different forms of geo-economic reasoning, which suggest that the United States' first security priorities in the post–Cold War era are domestic, not international (although what this means in a globalizing international economy is not clear). The election of President Clinton after he campaigned almost exclusively on the poor state of the U.S. domestic economy further eroded the sway of the classic Cold War geography of proximity and distance in U.S. strategic thinking. In his inaugural address, Clinton noted how "communications and commerce are global, investment is mobile, technology is almost magical, and ambition for a better life is now universal."[4] The new interdependent world of instantaneous informational, economic, and financial flows demanded a new cartographic consciousness within the United States. "There is no longer," Clinton asserted, "a clear division between what is foreign and what is domestic." This blurring of the inside and the outside of the nation, this melting of the internal and the international, made a new paradoxical foreign policy imperative possible. America's first foreign policy priority, Clinton declared, was domestic renewal.

The Clinton administration's attempt to redraw the governmental map of the United States' strategic responsibility and commitment has, however, been complicated by the power of global television images. Since 1989, American television audiences have been transfixed by dramatic geo-political spectacles: Tiananmen Square, the fall of the Berlin Wall, the release of Nelson Mandela, the Gulf War, the Soviet coup of August 1991, and the Yeltsin attack on the Russian parliament in the fall of 1993. All of these memorable spectacles were of limited duration and aesthetic televisual visibility.[5] Other political dramas, like the breakup of Yugoslavia, have had a longer duration and have not had the same transparency, visual aesthetics, and televisual intensity as instant blockbuster history. Nevertheless, a televisually slow event occasionally generates spectacular scenes worthy of projection into the living rooms and consciousness of global television audiences.

This televisualization and spectacularization of geopolitical conflicts throughout the globe is significant for established geographies of proximity and distance in that the immediate and instantaneous projection of remote conflicts can generate uncertain bonds of responsibility and detachment across television audiences worldwide. Global media vectors like television and telecommunications more generally make perception at a distance, what Mackenzie Wark calls *telethesia,* possible, thus creating a "virtual geography" that "doubles, troubles and generally permeates our experience of the space we experience firsthand."[6] When managed and screened properly, as happened during the Gulf War, television images from carefully directed and produced geopolitical spectacles can reinforce politically appropriate feelings of identity with a U.S.-orchestrated war machine and detachment from the mechanized killing produced by it. Official governmental maps of the world are, in other words, affirmed and territorialized in proper ways. In other cases, such as Bosnia and Rwanda, feelings of empathy and a broad sense of moral responsibility are made possible by the unmediated screening of the horrors of war and displacement. Official governmental maps of conflicts may become deterritorialized in improper ways and reterritorialized in inappropriate ways by global media vectors. This televisual destabilization of the state's official map of governmental responsibility can and does cause problems for foreign policy decision-makers in that it transgresses and disrupts the careful maps of obligation and responsibility they seek to operationalize and maintain. Thus, as Anthony Lake notes, there is a need to make distinctions, distinctions that contain the ambivalent, unbounded responsibility generated by television images of global conflict. These distinctions are renewed efforts to rechart official governmentalized geographies of responsibility and apathy, obligation and indifference, proximity and distance. In studying the production of these irrevocably geo-political distinctions in U.S. foreign policy discourse, we are studying the recharting of world order by the U.S. state, the production of new lines of geo-power in the age of virtual geography.

In this chapter I wish to investigate how this respacing of world order works in detail by examining the sets of distinctions U.S. foreign policy makers have publicly used to inscribe "Bosnia" as a place within the U.S. geo-political imagination.[7] In contemporary political discourse, "Bosnia" is one of a number of competing paradigmatic signs of post–Cold War world disorder, a place characterized by ethnic fragmentation, a bloody territorial war, and the uncontrolled flow of refugees. As a site of an uncontrollable irruption of violence and chaos, "Bosnia" is like many other disorderly places in the world. To U.S. foreign policy makers, therefore, evocation of "Bosnia" by the press and others prompts the standard response that the United States cannot and will not be the "world's policeman." By this script, "Bosnia" is a place like many others, beyond the United States' universe of strategic and military obligation. Rather than being a strategic site, "Bosnia" is, as Clinton remarked during the global forum, the site of a "humanitarian catastrophe" and, as such, requires a measured, circumscribed U.S. response befitting its status as such a site. However, as Christiane Amanpour's challenge to Clinton makes clear, the location of "Bosnia" within an official map of the United States' obligation and responsibility is an essentially contested one. As a location on the European continent, Bosnia is closer to an imaginary Western home than other locations, like Rwanda. More important, and like Rwanda also, Bosnia is more than simply the site of a "humanitarian catastrophe." As prominent journalists and many others have argued, "Bosnia" is the location of a modern-day "Holocaust" that is a consequence of the outbreak and spread of a Serb-sponsored fascistic dream of ethnically cleansed space and ethnically pure territory. As such, "Bosnia" demands a much more active, interventionist foreign policy response on the part of the United States.

The exchange between Clinton and Amanpour over the significance of "Bosnia" is symptomatic of a larger contest within U.S. political discourse, a contest over the writing of the meaning of "Bosnia" for U.S. foreign policy. As "Bosnia" emerged as a sign of post–Cold War chaos and ethnic hatred in 1991, two competing scripts struggled to enframe it within the U.S. geo-political imagination. Like all foreign policy scripts, each of these scripts was associated with a clear policy imperative for U.S. policymakers. Both these policy imperatives are sloganistically encapsulated by the same idiomatic declaration: "Never again." The first script, which was a script generated both within the global mass media and by the mid-level U.S. foreign policy diplomats who read the daily cables coming from the U.S. embassy in Belgrade, wrote "Bosnia" within the terms of a modified World War II script as the site of a modern-day "Holocaust." The Serb policy of "ethnic cleansing" that began in Krajina in Croatia in 1991 and spread to Bosnia in 1992 was interpreted as a horrific case of "genocide" once again irrupting on the European continent. The policy imperative for the United States, in the face of this genocide, was its officially enshrined attitude given the historical experience of the Holocaust: never again should the United States be a bystander while genocide unfolds.

The second script, which was the script first adopted by the foreign policy leaders in the Bush administration and later in the Clinton administration, wrote "Bosnia" within the terms of a World War I/ Vietnam script wherein "Bosnia" was a site that was a dangerous military quagmire. "Bosnia" was not a place of solid ground for U.S. policymakers; rather it was, as it was repeatedly described after 1991, a "morass," "vortex," "whirlpool," or, as Clinton noted during the global forum, a "sinkhole." All these feminized place descriptions—locations where excess fluids threaten to engulf solid forms—signaled danger.[8] The policy imperative they helped codify was a post-Vietnam "never again": never again should the United States get sucked into a civil war in a marginal place where the strategic stakes are murky and ill defined. For President Johnson, Vietnam was the bitch that pulled him from the woman he really loved: the Great Society.[9] This psychoanalytic register of the equivalence Vietnam–quagmire–castrating whore is an important one to bear in mind when discussing quagmires and quagmire anxiety.[10]

The operation of these two scripts wrote the imaginative geopolitical topography of "Bosnia" in quite different ways. In the first script, "Bosnia" is the site of a clear moral struggle between good and evil, victims and perpetrators, with the United States and other Western powers occupying the morally reprehensible role of "bystanders." The second script, however, accentuates the clouded, ambivalent, and confused nature of the Bosnian conflict. Whereas the first script foregrounded the moral clarity of the conflict, the second tended to emphasize its obscurity, noting that its origins stretch back thousands of years into ancient history.

Both scripts geo-politically produce "Bosnia." "Bosnia" is a site that is between a holocaust and a quagmire in the U.S. geopolitical imagination. This geo-political imagination comprises both an imaginary and a symbolic writing of "Bosnia," the former being an in-sight-ing of it within a geography of images (televisual, photographic, and cinematic), the latter being an in-site-ing of it within a geography of words, slogans, and narratives. Together they in-site "Bosnia" as a location for a new world spacing; they produce it as a holder of displaced and condensed signs, a World War I/World War II/Vietnam, a quagmire/holocaust. Unlike standard representations of the Soviet Union during the Cold War or Iraq during the Gulf War of 1991, Bosnia is not a one-dimensional place in the U.S. geopolitical imagination, the consistent home of the enemy or the Other. Bosnia, rather, is a place in-between, a place that immediately evokes a strategic, political, and moral dilemma for most members of the U.S. foreign policy establishment.

Given this, it is not surprising that U.S. policy toward this place has been viewed as having an oscillating "stop-go" or "flip-flop" quality about it, for, depending on the daily turn of the spectacle/story, "Bosnia" is capable of being both the site of a modern-day

Holocaust that morally demands active U.S. engagement and the site of a Vietnam-like swamp that threatens to suck the U.S. military into its vortex and immobilize it in a no-win situation. This dilemma is graphically illustrated in a cartoon by KAL from May 1994 in which "Bosnia" is depicted as both Europe during the Second World War and Indochina in the late 1960s. "Bosnia" is a place where the U.S. agent of global responsibility and obligation (Uncle Sam as a U.S. soldier) is hailed or interpellated as a witness to slaughter, yet is also a subject under threat, a subject who is sinking in a morass and in danger of dissolution (see Figure 12). If one interprets foreign policy as part of the process by which states seek to secure a positive ego ideal for themselves, then clearly "Bosnia" presents a conundrum for the U.S. subject and for U.S. subjectification. On the one hand, "Bosnia" is an opportunity for the U.S. subject to secure for itself its positive ideal as moral agent and rescuer of humanity, an ego ideal that is a deeply ingrained part of the mythology of American national exceptionalism. On the other hand, "Bosnia" endangers the self-image of the "United States" in that, if the latter intervenes, Bosnia threatens to envelop it in its uncertainty and indeterminacy; it is a potential density that emits no light, a nascent black hole that refuses to reflect a positive mirror image of the "United States" to itself. Alternatively, by not intervening, it jeopardizes the United States' "worldwide credibility" and risks undermining its image as a superpower, the very charge that Amanpour levels against Clinton.

In this chapter, I use official U.S. government policy statements and documents together with newspaper reports and discussions on U.S. policy toward "Bosnia" in order to trace the development of the geopolitical writing of "Bosnia" from 1991 to July 1994. (The role played by the United Nations and the European powers in the development of this script is not considered here in detail.) The first part examines the emergence of these scripts as the Bosnian crisis began under the Bush administration. The second part then considers their functioning during the Clinton administration. Part of my purpose is simply to document the story of the evolution of U.S. foreign policy thinking on Yugoslavia and Bosnia as the crisis developed. The record of this policy evolution bears out, I believe, the common charge that U.S. foreign policy toward Bosnia from the beginning was characterized by confusion, oscillation, and "flip-flops." Part of my purpose also, however, is to document the institutional forces

Figure 12. Bosnia: Between a holocaust and a quagmire. By kind permission of the Cartoonists and Writers Syndicate.

and personality dynamics that went into the codification of the two competing geo-political scripts producing "Bosnia" as a meaningful drama in U.S. foreign policy discourse. Whereas the "flip-flop" charge may be justified in policy development terms, it is not justified in geo-political terms, for U.S. foreign policy discourse since 1991 has consistently geo-graphed "Bosnia" as a potential quagmire that the United States must avoid. Although the holocaust script has had strong advocates within the U.S. government and foreign policy community, and although it has always shaped the con-textuality of "Bosnia" within this community, it has never been the dominant script. In the final analysis, "Bosnia" was written by the U.S. foreign policy community between 1991 and 1994 as a "faraway place," a place demanding "humanitarian aid" but ultimately a place beyond the United States' universe of obligation.

THE BUSH ADMINISTRATION AND BOSNIA

The Revolt against Centralism: 1990 to June 1991

In October 1990 a confidential CIA report predicted in a National Intelligence Estimate that Yugoslavia would break apart within eighteen months. The main culprit in this split, according to the report, would be Slobodan Milosevic. Given the frenetic diplomatic and military activity that was triggered by the Iraqi invasion of Kuwait that August, it is not surprising that the CIA report did not get the attention it deserved at the highest ranks of the U.S. government. The Bush administration policy toward the evolving crisis in Yugoslavia was in keeping with its policy toward Eastern Europe and the Soviet Union generally. Emphasizing stability over territorial restructuring, the Bush administration opposed the breakup of multinational states like Yugoslavia. This policy was influenced by Bush's determination not to complicate the efforts of then Soviet president Mikhail Gorbachev to keep the Soviet Union together.[11]

There were three geo-political principles underpinning U.S. policy in 1990 and the first half of 1991. The first of these was that the problems of Yugoslavia were an internal matter to Yugoslavia. In testimony before the Senate Foreign Relations Committee on February 21, 1991, Richard Schifter, assistant secretary for Human Rights and Humanitarian Affairs, chronicled the growing tensions and human rights abuses in Yugoslavia (particularly in Kosovo) but declared in conclusion that the "Yugoslavs themselves will have to set their house in order."[12]

The second principle was a commitment to a unitary Yugoslavia. Responding to the Croatian referendum vote for independence in May 1991, the U.S. State Department released a statement that described U.S. policy toward Yugoslavia as based on the interrelated objectives of democracy, dialogue, human rights, market reform, and unity. The most significant of these objectives, given the emergent power struggle in Yugoslavia, was the emphasis on unity, a principle that was elaborated in greater detail than the others: "By unity, we mean the territorial integrity of Yugoslavia within its present borders. We believe that the ethnic heterogeneity of most Yugoslav republics means that any dissolution of Yugoslavia is likely to exacerbate rather than solve ethnic tensions." The United States, the statement noted, "will not encourage or reward secession; it will respect any framework, federal, confederal, or other, on which the people of Yugoslavia peacefully and democratically decide. We firmly believe that Yugoslavia's external or internal borders should not be changed unless by peaceful consensual means."[13] Less than a month later, Secretary of State James Baker reiterated this position in Belgrade after meeting with Yugoslav leaders, stating categorically, in the face of expectations that Slovenia and Croatia were about to declare their independence, that "it would not be the policy of the United States to recognize that declaration, because we want to see this problem resolved through negotiation and through dialogue and not through preemptive unilateral actions."[14] Four days later, on June 25, 1991, Croatia and Slovenia nevertheless declared their independence.[15]

The third geo-political principle of U.S. foreign policy toward Yugoslavia at this time was that Yugoslavia was primarily a European problem, an emergent ethnic crisis in the backyard of the European powers. Because of its nonaligned status and nonmembership in the Warsaw Pact organization, Yugoslavia was never a primary Cold War locale. Its geo-political status, within the strategic vision of the North Atlantic Treaty Organization (NATO), was regional, not global.

The JNA Military Response: June 1991 to December 1991

All three of the geo-political principles underpinning U.S. foreign policy toward Yugoslavia in mid-1991 were overtaken by events and rendered obsolete. The first principle to go was the assumption that the crisis in Yugoslavia was an internal problem. In what some U.S. foreign policy officials now concede to be the case, the Baker visit to Belgrade ended up facilitating a military response by the Yugoslav army, the JNA, to the secessionist attempts in Slovakia and Croatia. After Baker's visit, significant fighting erupted in Yugoslavia. On September 2 the U.S. government called for a cease-fire in Yugoslavia and supported the establishment of a European Community (EC) conference on Yugoslavia under the chairmanship of Lord Carrington. On September 25 the UN Security Council (UNSC) passed an arms embargo against Yugoslavia. "The aggression within Yugoslavia," Secretary Baker declared, "represents a direct threat to international peace and security.... [N]o territorial gains or changes within Yugoslavia brought about by violence are acceptable."[16]

As the violence intensified, the second principle was also being questioned. Ralph Johnson, principal deputy assistant secretary for

European and Canadian Affairs, testifying before the Senate Foreign Relations Committee on October 17, stated that

> from the beginning, our fundamental policy objective in Yugoslavia
> has been democracy, not unity. But when the Yugoslav crisis began,
> we decided to state our support for both unity and democracy because
> we believed that unity offered the best prospects for democracy and
> stability throughout Yugoslavia. Given Yugoslavia's crazy-quilt eth-
> nic makeup and history of deep-seated ethnic disputes, we believed
> that the only alternative to some form of democratic unity was vio-
> lence, suffering, and long-term instability.... In light of [the violence]
> we have stopped talking about unity, not because we no longer think
> it the best solution, but because the deterioration of the situation has
> made other goals more immediate.[17]

Europe, Johnson added, "has the most at stake in this crisis."[18]

With the shelling of Dubrovnik and the siege of Vukovar in late
October and early November, the Yugoslav crisis was at a crucial
turning point. It was at this point that two complementary scripts
of the meaning of the evolving Yugoslav situation for the United
States were codified and came to rigidly designate the Bush administration's response to the crisis. Warren Zimmermann, the U.S. ambassador to Yugoslavia at the time (who later resigned from the Clinton
administration partly in protest over its Bosnia policy), identified
these as the "Vietnam Syndrome" and the "Powell Doctrine."[19]
The first script was assembled by the two highest-ranking experts
on Yugoslavia in the U.S. government: Deputy Secretary of State
Lawrence Eagleburger, who had served as ambassador to Yugoslavia
for seven years, and National Security Adviser General Brent Scowcroft, who also served for a considerable time in Belgrade with Eagleburger. Both men spoke some Serbo-Croatian and had left Kissinger
and Associates to take up their positions in the Bush administration, where they had previously lobbied on behalf of, among others,
the Yugoslav government. It was Eagleburger and Scowcroft who
began to speak of Yugoslavia as a potential "quagmire" for the
United States. Both Scowcroft and Eagleburger reportedly urged the
president to "stay away from Yugoslavia's tangle of ethnic passions,
warning of a diplomatic and military quagmire."[20]

The second script was codified at the Pentagon by General Colin
Powell and his advisers. The so-called Powell Doctrine has two dimensions. The first holds that the United States should avoid any

military involvements in which a mission statement, military goals, and an exit strategy are not clear from the outset. The United States should avoid "murkily defined missions" for its troops and pursue only those missions that are "doable." In the sloganistic terms of the Pentagon, "Without an exit strategy, a way to win, a way out, do not go in." The second dimension to the doctrine holds that if the United States is going to commit its military, it should do so with overwhelming force so as to secure a quick and easy victory, thus facilitating early withdrawal of U.S. troops. This second dimension is often described as an "all or nothing" strategy for using force, the best examples of its application being the Panama invasion and the Persian Gulf War of 1991.

In applying the Powell Doctrine to the situation in Yugoslavia, the Pentagon's military planners decided that Yugoslavia was precisely the type of situation where the United States should not intervene militarily. Like the Eagleburger-Scowcroft reading, it too designated Yugoslavia as a dangerous quagmire that the United States should avoid. Not all of the U.S. foreign policy planning staff agreed, however. In particular, many of the junior- and middle-level desk officers in the State Department believed that the United States should have taken a stronger position against "ethnic cleansing," a translation of the term the Serbs used to describe their policy in Bosnia and Croatia at this time. One of the first of a number of desk officers to resign in protest was George Kenney. He described Colin Powell as always a strong voice against doing anything.[21]

The upshot of this relative consensus of the elite within the Bush administration on Bosnia in late 1991 was that the United States' only response to the attempt to create an ethnically pure "Greater Serbia" was an arms embargo and economic sanctions against all six provinces. According to Warren Zimmermann, Yugoslavia had become a "tar baby" in Washington by late 1991. "Nobody wanted to touch it.... [I]t was seen as a loser."[22] The Bush administration committed itself to a policy of doing nothing in a substantive military way to check the forceful implementation of a fascistic vision of an ethnically pure state on the continent of Europe. Reflecting on the policymaking of Bush and Baker at this time, one close observer remarked that the two Texans "thought it was a swamp."[23] Although there was in actuality considerable policy consensus at the top at the time, Eagleburger later recalled that "we didn't have an agreement among ourselves on how important it was, how dangerous it was, and by the time it got dangerous there were these splits within the Western community."[24]

Recognizing Newly Independent States:
December 1991 to April 1992

The splits Eagleburger was referring to were those emerging both within the European Community and between it and the United States. The sudden activism of Germany in pushing for the recognition of Croatia and Slovenia, its traditional allies during World War II, reportedly startled Washington.[25] On December 23, 1991, Germany, after making its intention to do so clear a few months earlier, recognized Slovenia and Croatia as independent states. On January 15, 1992, the other members of the European Community followed Germany's lead. The U.S. government also began to reconsider its policy on recognition, helped by the international outcry at Serb aggression against Croatian towns and villages. In Belgrade, U.S. ambassador Warren Zimmermann favored the recognition of Bosnia as a means of keeping the JNA's war of aggression from spreading to Bosnia. He later stated that "our view was that we might be able to head off a Serbian power grab by internationalizing the problem. Our hope was the Serbs would hold off if it was clear Bosnia had the recognition of Western countries. It turned out we were wrong."[26]

The American plan to recognize an independent multiethnic Bosnia proceeded alongside European efforts to facilitate the ethnic division of Bosnia. In February 1992 Radavan Karadzic, Mate Boban, and Alija Izetbegovic had agreed to partition Bosnia after European Community–sponsored talks in Lisbon, but, in circumstances that are unclear, Izetbegovic backed away from the plan.[27] On February 26, in testimony before the Senate, Secretary of State James Baker declared that "we are reviewing our recognition policy on the Yugoslav republics on an almost daily basis, or certainly weekly basis."[28] After a Serb-boycotted referendum in late February, Bosnia-Herzegovina declared its independence. Secretary of State Baker met with EC leaders and urged them to recognize Izetbegovic's government. Baker told the Europeans to stop pushing the "ethnic canonization of Bosnia."[29] With the Serbs threatening civil war, Izetbegovic once again agreed to divide Bosnia into "three constituent units," only to back away from it in a few days, saying he only signed because Europeans were forcing him to do so. In the end, American efforts prevailed. On April 6, EC foreign ministers recognized the independence of the unitary state of Bosnia-Herzegovina. The next day the Bush administration recognized the independence of Croatia, Slovenia, and Bosnia (but not Macedonia). By this time, however, the JNA and Serb irregular invasion of Bosnia had begun. By April 21 Serb forces were bombarding Sarajevo.

According to Zimmermann, the Serb dismemberment of Bosnia was made easier by EC irresponsibility, U.S. passivity, and the miscalculations of Bosnian president Izetbegovic. But it has also been claimed that the United States made a fundamental mistake in pushing for the recognition of a single independent Bosnia. This mistake was not simply in recognizing an independent multiethnic Bosnia but in not doing anything to protect the territorial integrity of this state once it came under attack. U.S. foreign policy leaders intervened decisively with their European colleagues to make the creation of a single Bosnian state possible while simultaneously holding to the attitude that Yugoslavia as a whole was a European problem and a potential "quagmire." Bosnia was a relatively low priority for the U.S. foreign policy apparatus—the U.S. State Department only began a chronology of Bosnian political developments in March 1992— and unity with the Europeans would instinctively come before any major policy rupture over Bosnia.

Watching "Ethnic Cleansing": May to December 1992

By recognizing but not militarily supporting an independent Bosnia, the United States was acknowledging a place that became the site of the most brutal war witnessed on the European continent since World War II. Out of the daily transmission of images of "ethnic cleansing," stories of mass rape, and, later, the discovery of concentration camps emerged a powerful mass media script of Bosnia as the site of a modern-day genocide. On May 27, a mortar attack on bread lines in Sarajevo killed twenty people and wounded seventy. Television cameras recorded the carnage and projected it into living rooms around the world. The United States responded by pushing new economic sanctions (UN Resolution 757) against Serbia and Montenegro through the UN Security Council (UNSC) in three days. However, while U.S. policymakers were not inactive on Bosnia, they now had a perverse interest in making sure that Bosnia as a place was not read as the site of a new holocaust. When the first reports of Serb-run concentration camps reached Washington in May 1992, the administration delayed an investigation, fearing it would step up pressures for intervention.[30] Bosnia was, for the Bush administration, a potential morass before anything else. Only after stating this was it possible to acknowledge it as a "humanitarian nightmare," a Baker description that wrote place in terms of outcome, not process. Encapsulating this attitude, Bush stated on July 10: "I think people are reluctant to get bogged down in a kind of guerilla warfare. And I also think that the main objective, now, of humanitarian relief is ... the key thing."[31]

The pictures of emaciated captives being held in detention camps in Omarska, Bosnia, by Serbs on August 6, however, made containing the holocaust script extremely difficult.[32] U.S. presidential candidate Bill Clinton and his running mate Al Gore called for air strikes against the Serbs. President Bush responded with his first major statement on the situation in the former Yugoslavia. He described the war there as a "complex, convoluted conflict that grows out [of] age-old animosities. The blood of innocents is being spilled over century-old feuds." Read symptomatically, Bosnia was a place of spillage for Western leaders like Bush, a place of uncontrolled flows that required "containment" (the very title of Bush's speech — "Containing the Crisis in Bosnia and the Former Yugoslavia" — is symptomatic of this desire). In contrast to the Gulf Crisis, where the United States could draw a clear "line in the sand," "the lines between enemies and even friends are jumbled and fragmented." Unlike the deserts of the Middle East, Bosnia's physical and also moral topography was mountainous, dense, impenetrable, and murky. Bosnia seemed to suggest a classic American image of clan feuds among isolated mountain peoples. "Blood feuds" Bush declared, "are very difficult to resolve."[33]

The reading of the situation in the former Yugoslavia as a "blood feud" was not, at first glance, unreasonable. This and the complex, age-old animosity descriptions performed a very important geographing function in distancing the wars in Bosnia and Croatia from the West's own self-image. The West lives in a modern present and not in the past. Its space is that of modern, rational civilization and not the traditional primitivism of "blood feud" cultures. The Bosnian war was, in one sense, not European, but an irruption of a premodern or Orientalist/Africanist/Slavic savagery (a white man's Rwanda) in the heart of Europe, the continent of freedom, enlightenment, and the rule of law.[34] The situation in the former Yugoslavia is the return of the past and the primitive, to what the Bush administration frequently described as the "law of the jungle." The new acting secretary of state Lawrence Eagleburger declared that the "civilized world simply cannot afford to allow this cancer in the heart of Europe to flourish, much less spread. We must wrest control of the future from those who would drag us back into the past."[35]

This type of reasoning is significant in that the conflict is not understood as both a modern and a rational war by former Communist institutions and elites to try to preserve and strengthen their power in a post-Communist world.[36] Serbian aggression was represented as an irruption of barbarism, an irruption that was obviously distressing ("a true humanitarian nightmare," in Bush's terms) but not, because of its attributed Orientalist/Africanist/Slavic primitiveness, within the same moral universe and community of obligation as "us," the truly modern, rational, and civilized. The wars in the former Yugoslavia were taking place in Europe, but they were not of Europe. David Rieff sums up this cartographic vision: "Europeans could not do these things to one another, therefore the inhabitants of the former Yugoslavia could not be Europeans." The governmental map of the Bosnian slaughter that developed was "based on the most old-fashioned, essential notions about the Balkan character, and came accompanied with much pseudohistorical guff about ancestral hatreds and a regional predisposition to violence. It was a story that effectively read all the South Slavs with the exception of the Slovenians out of Europe. Europeans didn't act that way—not real Europeans, anyway."[37]

In responding to questions after his major statement, Bush evoked the imagery of a Vietnam-like quagmire, declaring, "I do not want to see the United States bogged down in any way into some guerilla war—we lived through that once."[38] This case against any form of U.S. military action in Bosnia was forcefully promoted in both private and public by General Colin Powell. Responding to international condemnation of U.S. policy from, among others, Margaret Thatcher, Jeane Kirkpatrick, and George Schultz, Powell conducted a campaign through the media against what he termed "the impetu-

204 · BETWEEN A HOLOCAUST AND A QUAGMIRE

ousness of civilians," who he claimed were too quick to place American forces in jeopardy unwisely and for ill-defined missions.[39] Powell's classic example of such a case was Beirut in 1983, where marines were put in "as a symbol, as a sign." Because they did not know what their mission was, because the situation was confusing, according to Powell, 241 of them died. Powell opposed a ban on Serbian combat flights over Bosnia within the administration, but President Bush eventually decided to support such a ban on October 12.[40] Continued media criticism of the Powell Doctrine prompted Powell to write (in an unprecedented move) an opinion editorial piece in the *New York Times* in which he described the crisis in Bosnia as "especially complex...one with deep ethnic and religious roots that go back a thousand years." President Bush, Powell argued, more than any other recent president, understood the proper use of military force. "In every instance, he has made sure that the objective was clear and that we knew what we were getting into. We owe it to the men and women who go in harm's way to make sure that their lives are not squandered for unclear purposes."[41]

The murkiness and the unclear nature of the Balkans conflict were, however, not immanent features of the war itself. By describing the conflict in Bosnia as having thousand-year-old roots and making the decision not to explain it in the moral terms used, for example, during the Persian Gulf crisis, the Bush administration was effectively contributing to the murkiness and mystique of the Balkans as a region. Because of their historical significance as a place of Great Power contestation in the nineteenth century and the place where World War I began in 1914, the Balkans were a condensed signifier of endless strife, endemic violence, and war. The Balkans were the homeland of ethnic conflict, a region not worth, in Bismarck's oft-quoted description, "the life of one single healthy Pomeranian grenadier."

The question of "Bosnia" was not a significant issue during the presidential election of 1992. Only after the election did Secretary of State Lawrence Eagleburger begin to emphasize the humanitarian dimensions of the Bosnian situation. In mid-December Eagleburger had a talk with Holocaust survivor and Nobel Peace Prize laureate Elie Wiesel, who supposedly impressed upon him the need to understand "Bosnia" as a "humanitarian tragedy." Eagleburger explained the encounter thus:

The fact of the matter is that things are going on there that are absolutely outrageous of which ethnic cleansing is sort of the catch-all phrase for this but, prison camps, torture, and all of those things. So, I decided, after listening to Elie Wiesel, that he was probably right. We needed to take this to a different level. I don't know that it's going to solve anything or change anything, but I think it's time, when we have the facts—and we do in some of those cases fairly clearly—that we begin to name names.... [F]or whatever period that is left to us, that is until the 20th of January, this is going to be a theme. And I hope the next administration will pick it up.[42]

Eagleburger's speech to the International Conference on the Former Yugoslavia in Geneva on December 16 marked the beginning of this belated "naming names" strategy, a strategy that embedded events in the former Yugoslavia within a holocaust script to a greater extent than any previous descriptive strategy of the administration. He reaffirmed the U.S. commitment to a unitary Bosnia, declaring, "We will settle for nothing less than the restoration of the independent state of Bosnia-Herzegovina with its territory undivided and intact, the return of all refugees to their homes and villages, and indeed, a day of reckoning for those found guilty of crimes against humanity."[43] On December 17 NATO foreign ministers supported enforcement of the no-fly-zone resolution (Resolution 781) that the Security Council had adopted on October 9. It also committed itself to the proposition that the "sovereignty and territorial integrity of Bosnia-Herzegovina must be restored."[44] Meanwhile President Bush wrote a letter to President Milosevic and the Yugoslav army chief General Zivota Panic warning that the United States was now ready to use force if UN forces were attacked in Bosnia-Herzegovina or if relief efforts were hindered. The letter went on to declare that "in the event of conflict in Kosovo caused by Serbian action, the United States will be prepared to employ military force against the Serbs in Kosovo and in Serbia proper."[45] The Bush administration, at the end of its tenure, was belatedly recognizing that events in the former Yugoslavia might end up requiring a military response by the United States.

CLINTON AND BOSNIA

A "New" Policy: January to March 1993

Hopes were high among pro-Bosnia groups that the new president would reverse the Bush administration's failure to demonstrate effective military leadership of the Western states on the Balkans crisis. During the campaign candidate Clinton, reacting to the revelation of concentration camps, had called for the use of "air power against the Serbs to try to restore the basic conditions of humanity" in Bosnia.[46] However, while the new Clinton administration promised a more resolute policy against Serbian "ethnic cleansing," candidate Clinton was nevertheless a reluctant foreign policy leader and someone who concurred with the view that Bosnia was a quagmire. During the campaign (August 11, 1992), Clinton stated that "we cannot commit ground forces to become involved in the quagmire of Bosnia."[47] Clinton's predisposition to view Bosnia as a potential quagmire was strongly reinforced in a series of preinauguration briefings he received from General Colin Powell and the Pentagon.[48]

The wisdom of the Powell Doctrine was questioned at this time in the press by the British journalist Martin Walker, who charged that it underpinned a "geopolitics of casual sex," or the promiscuous and irresponsible use of U.S. military force without lasting commitment.[49] Walker highlighted the explicitly geographical dimensions of the Powell Doctrine, albeit in an exaggerated way. Only in places where the United States can expect a quick win shall overwhelming strength be used. In the language of one late-night comic: "We do deserts; we don't do jungles. Or mountains." Walker argued that the United States' spasmodic international invasions in Panama, Kuwait, and Somalia, together with lesser deployments in the Philippines and Liberia, constituted examples of this "geopolitics of casual sex," a geopolitics, Walker argued, that the United States needs to stop if it is to play a constructive, responsible, and stabilizing role (presumably with "staying power") in the post–Cold War era. The register of Walker's argument is interesting in two ways. First, his argument implicitly understands the end of the Cold War as an era characterized by the loss of a responsible father figure. The U.S. state is suffering from declinism; its foreign policy has become narcissistic; the world is without a steady guardian. Walker's argument resonates with a certain anxiety within the corridors of power in the Pentagon, a certain post–Cold War culture of doubt and ebbing self-confidence in America's previously unchallenged position in the world. Second, Walker identifies an impatient and insecure masculinist military gaze operating in U.S. foreign policy. It sees certain sites as appropriate places for necessary demonstrations of military prowess in world affairs and certain sites as locations of danger and potential engulfment. The desert *theater* is the most appropriate place for the demonstration of prowess in that everything is visible and controllable, a perfect platform for a high-technology spectacle of dazzling brilliance, a splendid stage for the simulation of a depthless "new world order."[50] By contrast, jungles and mountains are spaces of occluded vision and unknown depth, feminized regions of potential embarrassment that induce hyperanxiety and performance failure.

Once the Clinton administration took office, the new secretary of state, Warren Christopher, ordered a rush review of U.S. policy toward the Balkans. Christopher saw it as his bureaucratic job to keep foreign policy from getting in the way of Clinton's domestic agenda. According to Elizabeth Drew's portrait of foreign policy decision-making in the Clinton White House, the legalistic Christopher was not comfortable dealing with the Balkans because it could not be approached country by country. In the Balkans there were no countries in the usual sense, so it defied "a rational or linear approach."[51] Resulting from the policy review were six new policy steps, presented, significantly, not by the president but by Secretary Christopher in a speech on February 10, 1993. First, the Clinton administration decided to back the Vance-Owen negotiations and appointed an envoy to the talks. Second, the administration backed a negotiated solution. Both these steps, in effect, meant that the Clinton administration was abandoning the earlier Bush administration policy of supporting a multiethnic single Bosnian state. It also effectively meant that the United States was giving up its principled opposition to a proposed ten-province carve-up of Bosnia (the Vance-Owen map) that rewarded Serbian "ethnic cleansing." Third, the administration promised to tighten the sanctions against Serbia. Fourth, it committed itself to enforcing the no-fly zone over Bosnia, to further humanitarian aid, and to the opening of a war crimes tribunal (approved by the UNSC on February 22). Fifth, it promised the possibility of U.S. military participation to implement and enforce any viable agreement between the parties. Sixth, it promised to work closely with Russia in the search for a negotiated solution.[52]

By backing the Vance-Owen talks, the Clinton administration effectively relinquished its independent-initiative capacity and turned responsibility for the conflict over to the Europeans and the United

Nations. "The United States," Christopher declared, "is not the world's policeman. We cannot interpose our forces to stop every armed conflict in the world." Echoing Christopher's wariness of getting involved in such a notorious twentieth-century flash point, Clinton remarked the same day, "It's no accident that World War I started in this area. There are ancient ethnic hatreds that have consumed people and led to horrible abuses." Bosnia for Clinton was "over there," a distant place that threatened to pull the United States "into it in horrible ways" if the United States did not follow the right policy.[53] Read symptomatically, Clinton was defining his states-manhood by his will to resist the demanding bitch of a problem that was Bosnia.

Geopolitically the Clinton policy review reinscribed Bosnia once more as a European problem, a dangerous vortex that was the site of a "humanitarian problem" but also a site where the U.S. military should not be involved. Specifically, the United States' involvement was not to be on the ground in Bosnia but in the air. The Clinton administration concentrated its efforts in February on coordinating a two-track humanitarian mission in Bosnia with the Europeans. The French, British, and other powers were to provide aid by land, while the United States committed itself to deliver aid by air to areas that could not be reached by land.[54] This geographical division of labor was an expression of the political parameters the United States and the Europeans set for themselves in the Balkans. Since the Balkans were in Europe, the Europeans should be "engaged" on the ground. Since the United States was committed to the security of Europe in a general, abstract way but had pressing domestic priorities, the Americans would limit their engagement to aerial work. Both engagements were defined by the politically palatable but practically vague and, as the conflict progressed, increasingly unworkable notion of a "humanitarian mission."

A first indication, to the Clinton administration at least, of the contradictions of the West's "humanitarian mission" was the fall of the besieged Muslim town of Cerska on March 2 to the Serbs, a town where the Americans had only begun dropping food and medicine. Asked to respond to the charge that U.S. aid drops were actually increasing violence and ethnic cleansing in Bosnia, Clinton responded by reiterating that he would not "commit the United States to a quagmire" and would not introduce U.S. ground troops.[55] The signing of the Vance-Owen plan by Alija Izetbegovic in March appeared to offer a sign of progress for U.S. policy. Also on March 31, the UNSC stepped up the pressure on the Serbs by endorsing the use of force to patrol the no-fly zone over Bosnia. However, the Bosnian Serb refusal to sign the Vance-Owen plan, the siege of Srebrenica, and the persistence of stark images of ethnic cleansing from Bosnia left the Clinton administration's policy adrift. Clinton himself stated that these events left him "outraged," but he and his Principals Committee (his chief foreign policy advisers) struggled to formulate a new policy in response.[56]

Holocaust and Balkan Ghosts: April–June 1993

On April 22, the U.S. Holocaust Museum was dedicated in an international ceremony. In a State Department lunch to honor Elie Wiesel, the key figure behind the museum, Richard Johnson (head of the State Department's Yugoslavia desk from 1990 to 1992) wrote that Wiesel argued that the Serb's conduct created a "moral imperative" for American intervention. Undersecretary for Political Affairs Peter Tarnoff responded that failure in Bosnia would destroy the Clinton presidency and that there were higher moral stakes than those in Bosnia, namely "the survival of the fragile liberal coalition represented by this Presidency." Johnson's report of this conversation later appeared in an unpublished paper entitled "The Pin-Stripe Approach to Genocide," which accused the president, the secretary of state, and senior State Department officials of playing down evidence that "ethnic cleansing" constituted genocide under international law.[57] As in the Bush administration, an "emotional" holocaust reading of "Bosnia" had to be contained. But, as Johnson's unpublished paper shows, there was dissension within the Clinton administration over Bosnian policy. On April 23, UN ambassador Madeleine Albright handed Clinton a letter urging air strikes against the Serbs, while twelve midlevel State Department officials also submitted a letter urging the president to use military force in Bosnia.

It was thus more than ironic that President Clinton, in his address on the dedication of the Holocaust Museum, noted how the Holocaust "reminds us forever that knowledge divorced from values can only serve to deepen the human nightmare, that a head without a heart is not humanity."[58] Clinton mentioned "ethnic cleansing" in the former Yugoslavia but described it as an example of "animosity and violence" and not in explicit holocaust terms. Elie Wiesel, in his remarks, turned to President Clinton and drew a comparison to the current events in Bosnia, declaring, "We must do something to stop the bloodshed in that country." Addressing this very comparison, the president himself said, "The Holocaust is on a whole different level."[59] Nevertheless, in late April Secretary of State Christopher met with Wiesel, who was becoming something of a conscience doctor to the foreign policy establishment, to discuss Bosnia's "moral implications."

On the day of the museum dedication, two leading members of the Senate Armed Services Committee, Senator John McCain (R-Arizona) and Senator Joseph Lieberman (D-Connecticut) debated Bosnian policy on CNN's news interview program *Crossfire,* with the plea by Wiesel as the lead into the debate. McCain described the conflict as a "centuries-old, since-the-Middle-Ages, ethnic conflict" and argued, citing (in his standard way) "military experts," against military air strikes. "It's a European problem to a large degree," and "the Europeans, if we contemplate ground troops, should be the ones in there, not American troops."[60] McCain essentially placed Bosnia beyond the United States' realm of military responsibility, although he did envisage the United States providing air support for European troops if it was decided that "ethnic cleansing" had to be stopped. Echoing this post–Cold War rewrite of the space of U.S. responsibility, the former Republican presidential candidate and *Crossfire* cohost Pat Buchanan stated that Bosnia is "clearly in Europe's backyard. Bosnia is as close to Berlin, Paris, and Rome as probably—certainly—Mexico City is almost to Texas." The problem is that the Europeans do not want to put their ground troops at risk. They do not want to put any men into battle. Cohost Mike Kinnsey concurred, suggesting that "the Europeans unfortunately don't seem to have much guts."

In response, Senator Lieberman cast Bosnia as a question of American identity, a moral identity without borders or limits. He argued that the United States is the world's only superpower and "one of the reasons we are is because we stand for what's right." Pointing to the Holocaust Museum dedication, he added that "the most moving movement was when that survivor of the concentration camp thanked and embraced the American soldier, the veteran, who liberated that camp. That's what we're about," he declared.[61] Other figures, from liberals like Senator Joseph Biden (D-Delaware) and Representative Frank McCloskey (D-Indiana) to neoconservatives like Jeane Kirkpatrick and Richard Pearle, made a similar case for U.S. military intervention in the region, in opposition to figures like McCain, Eagleburger, and Scowcroft.[62] Summarizing the policy debate in the Congress as a dilemma of competing analogical spaces, neoconservative Representative Henry Hyde (R-Illinois) declared that "people are trying to figure out whether this is Germany, 1942, or Vietnam, 1975." Biden made a similar point in noting that "lawmakers are haunted by the specter of Neville Chamberlain or the specter of Vietnam."[63]

The UNSC tightened sanctions against Serbia and Montenegro on April 17; after a short delay to give the Bosnian Serbs the possibility of accepting the Vance-Owen plan, they went into effect on April 26. The UNSC also declared Srebrenica and five other towns "safe areas, free from armed attacks and from any other hostile acts which endanger the well-being and the safety of their inhabitants."[64] Significantly, the besieged town was not termed a "safe haven" because, by precedence in northern Iraq and elsewhere, calling it a "safe haven" might have obliged the member states to guarantee the safety of the people living there. "Safe areas," therefore, was a bureaucratic category produced by a policy that read Bosnia only as a site of a humanitarian problem. It was a product of a paradoxical moral geography that included Bosnia within a general universe of humanitarian obligation but outside, for the moment, a practical universe of geopolitical responsibility and military action. Bosnia was governmentalized and mapped as a humanitarian and not as a genocidal problem.

On April 30 the Principals Committee met to finally establish direction in U.S. foreign policy in Bosnia, given the persistence of "ethnic cleansing." Lake, Gore, Albright, and Aspin all reportedly favored a more vigorous U.S. policy, while CIA director James Woolsey and Powell were against further U.S. involvement. The direction approved included a number of specific recommendations, including military air strikes.[65] Summarizing the decision reached, one top policymaker quoted by Drew remarked: "The basic strategy was, This thing is a no-winner, it's going to be a quagmire. Let's not make it our quagmire. That's what lift the arms embargo, and the limited air strikes, was about."[66]

Christopher went to Europe in the first week of May to seek approval for a U.S. plan to lift the UN arms embargo against the Bosnian Muslims and to threaten air strikes against the Serbs. The plan went by the shorthand "lift and strike." But Christopher went in what he termed a "listening mode" and not in the traditional leadership mode that had characterized U.S. relations with Europe during so much of the Cold War. This "listening mode" was considered appropriate because certain European powers (principally the French and the British) had their troops on the ground in Bosnia and because the Clinton administration still conceptualized the Balkans as primarily a European problem. Testifying before Congress upon his return, Christopher stated that "any intervention in such a morass must be carefully considered" and that "at heart this is a European problem."[67] In an unusually frank admission, Christopher declared that Bosnia was "a problem from hell."[68]

The result of Christopher's trip to Europe was a major embarrassment for the Clinton administration, as the Europeans publicly rebuffed both "lift" and "strike." The *New York Times* reported an off-the-record comment at the May 10 EC foreign ministers meeting suggesting that President Clinton "feels the compulsion to do something but does not actually want to get involved."[69] From the information that exists on the foreign policy decision-making process on Bosnia within the Clinton administration, this seems to have been the case. After the compulsion to act brought on by his visit to the Holocaust Museum, Clinton reportedly read Robert Kaplan's book *Balkan Ghosts,* which portrayed the region's peoples as having a long historic propensity for war with each other (General Powell and Hillary Rodham Clinton had also read the book), and began to back away from even his own "lift and strike" policy.[70] The result was foreign policy confusion within the administration for a time, although all within an agreed imperative not to "Americanize" the Bosnian question.[71]

The only genuinely new proposals to come out of the Joint Action Program announced by Christopher on May 22 after his European trip was a commitment to extend the mandate of the UN Protection Force (UNPROFOR) to deter attacks against the declared "safe areas" in Bosnia (approved in UNSC Resolution 836, June 3) to an increased international presence in Macedonia, and a statement to the effect that "the United States is prepared to meet its commitments to help protect United Nations forces in the event they are attacked."[72] This decision marked a decisive failure of political and military will. Rather than risk intervening militarily in the conflict, the United States placed greater emphasis on reestablishing Western unity and continued to governmentalize the crisis in Bosnia as a "humanitarian" one rather than a moral, political, and strategic one. Asked whether he thought the program would have any impact, President Clinton replied, "It could, it could," adding, "At least we [the Allies] are together again." Americans, he stated, "should be reassured that," as a result of the accord, "there is [only] a limited possibility of a quagmire" in Bosnia.[73] Reflecting on the failure of Western diplomacy in Bosnia at the end of May, an *Economist* editorial couched the issue in terms of geographical and moral distance and proximity: "The failure with ex-Yugoslavia is a failure of understanding. Most people in America and Western Europe still see what is going on there—no farther away from this summer's holiday-makers in Venice or Tyrol than Lyons is from Paris, or York from London—as a largely irrelevant struggle in an obscure corner of the world. It is not. It matters to the democracies because it is a battle about the ideas they believe in, because it is on their doorstep, and because the rest of the world has been watching to see what they would do about it."[74] But what the Clinton administration decided to do was to avoid Bosnia as a policy issue. In the following weeks, Secretary of State Christopher reportedly moved to "methodically shut down the Bosnia policy. He wanted to get the subject off the front pages."[75]

Learning to Say No: June–December, 1993

By choosing to devote its energies to establish so-called safe areas and then refusing to act when they were attacked or access to them was blocked, the Clinton administration had effectively decided that Bosnia was indeed, as Czechoslovakia was to Chamberlain, a "faraway place" that was not of fundamental strategic importance to the United States. The result was that U.S. policy for the next eight months was, in practical terms, committed to containing the televisually disseminated, emotive sign value of "Bosnia."[76] The administration did press ahead with "lift and strike," but the "lift" part went down to defeat in the UNSC on June 30. The "strike" part was slowly adopted by NATO as the Sarajevo siege got worse, and was finally adopted by the North American Council in Brussels on August 2.[77]

In September, the Clinton administration publicly elaborated its foreign policy philosophy in a series of keynote speeches. Anthony Lake, one of the more forceful voices on Bosnia within the administration, outlined a strategy of "enlargement" as the successor strategy to "containment" in the post–Cold War world. He noted that the conflict in Bosnia deserved American engagement, describing it as "a vast humanitarian tragedy" that can "all too easily explode into a wider Balkan conflict."[78] Within Lake's four components of the strategy of enlargement, Bosnia was predominantly part of the last, the "humanitarian agenda," the containment of chaos and the easing of suffering in places of "humanitarian concern." Public pressure for U.S. humanitarian interventions was driven by televised images, images that were implied to be fleeting and arbitrary in that they "depend on such considerations as where CNN sends its camera crews." Articulating a personal as well as strategic unease with the uncontrollable implications of television images, Lake argued that the United States needed to bring other considerations to bear as well: "cost, feasibility, the permanence of the improvement our assistance will bring, the willingness of regional and international leaders to do their part, and the likelihood that our actions will generate broader security benefits for the people and the region in question."[79] This policy need to develop criteria for U.S. humanitarian intervention was also emphasized by Albright, who outlined the criteria the United States would use when deciding to support a UN peacekeeping or peacemaking resolution. President Clinton played the same language game in his keynote address to the United Nations, making it clear that UN interventions had to be selective: "If the American people are to say yes to UN peacekeeping," he stated, "the United Nations must know when to say no."[80] In Bosnia, UN peacekeepers have maintained a valiant humanitarian effort, he said, while in Somalia "the United States and the United Nations have worked together to achieve a stunning humanitarian rescue" (651).

The fear that the spillage and unbounded flows associated with the "humanitarian agenda" would disable, immobilize, and suck the United States into dreaded quagmires was given new prominence a few weeks later. When the U.S. Rangers in Mogadishu, Somalia, got into trouble in October, a castration anxiety discourse of quagmire phobia developed in the Pentagon, the White House, and Congress. Because of the deaths of twelve soldiers, the televisual images of a captured and bruised American hostage, and the body of a U.S. soldier being dragged by a mob through the streets of Mogadishu, Somalia suddenly became transformed from the site of a "humanitarian mission" to what was represented as a full-blown "quagmire." The whole American body politic seemed under attack. Clinton was forced into a damage control strategy. He immediately bolstered the U.S. contingent of troops in Somalia, sharply curtailed their mission, and scheduled a firm date for their withdrawal. This policy was straight from the script of the Powell Doctrine: overwhelming force, a clear mission, and an established exit strategy. The "lesson" of Somalia for the Clinton administration was to avoid what the Pentagon began calling "mission creep" in UN peacekeeping work. This remarkable writing of Somalia as a "foreign policy failure," a failure that led to calls for Warren Christopher's resignation and finally contributed to the resignation of Secretary of Defense Les Aspin, further contributed to U.S. foreign policy drift on Bosnia. Yet more State Department officials resigned in protest over Clinton's policy on Bosnia. Marshall Harris, the State Department's desk officer for Bosnia from February through August, resigned in protest in September, while Warren Zimmermann, the former ambassador to Yugoslavia during the Bush administration, resigned in January 1994. When Christopher outlined the strategic priorities of U.S. foreign policy before the Senate in November, ending the war in Bosnia was not on his list. He had a different concern. In a proclamation illustrative of the contest over the mapping of Bosnia between state and media (and also between discourse and image), television images, he declared, "cannot be the North Star of America's foreign policy."[81]

Sarajevo and Gorazde: February to May 1994

Yet television pictures of the sixty-eight people blown apart and the two hundred people wounded by the mortar attack on a Sarajevo marketplace on February 5 were the immediate occasion for a reassessment and renewal of U.S. Bosnian policy. The bloody persistence of the war, the erosion of NATO credibility, and a Madeleine Albright report on the refugee crisis generated by the war were already provoking such a reassessment.[82] In announcing a French-inspired but Clinton-led NATO ultimatum to the Serbs to withdraw their heavy weapons behind a twenty-kilometer radius from the center of Sarajevo, President Clinton described U.S. interests not in terms of stopping the war but in terms of "preventing a broader conflict in Europe" and "showing that NATO, history's greatest military alliance, remains a credible force for peace in post–Cold War Europe."[83] The president also highlighted "the destabilizing flows of refugees" as a U.S. interest. Only last did he mention preventing the "strangulation" of Sarajevo and the "continuing slaughter of innocents" in Bosnia as a U.S. interest.

The success of the NATO ultimatum provided an impetus for the renewed diplomatic efforts of the U.S. government to find a settlement to the war. Under U.S. sponsorship, the Bosnian Muslims and Croats signed a peace agreement on March 18, endorsing, first, a constitution for a new bicommunal federation in Bosnia and, second, a preliminary agreement on a confederation between this federation and Croatia.[84] With the U.S. government firmly leading the search for a peace settlement and with decisive NATO action in late February downing four Serbian planes violating the no-fly zone, it seemed that an activist U.S. policy had, at last, materialized. However, momentum was lost in the policy debacle over Gorazde when various administration officials gave mixed signals as to the United States' commitment to the "safe area" of Gorazde, which was under renewed Serbian attack. After a week of remarkable oscillation, NATO planes lightly bombed Serb positions in Gorazde, ostensibly in response to attacks on UN personnel in the city. The Serbs, however, continued their offensive against the city's Muslim inhabitants, while the Clinton administration and NATO frantically sought for a way to shore up their credibility. The resulting second ultimatum, issued to the Serbs on April 22 by the North Atlantic Council (NATO's governing body), set a series of deadlines for the Serbs to withdraw from around Gorazde, all of which were violated, although no air strikes were undertaken. The relative slowness of U.S., UN, and NATO action in saving Gorazde's population from shelling at the hands of the Serbs was indicative of the priorities of U.S. interests.[85] Gorazde was significant as a place not because it was the site of a slaughter, another scene in the drama of ethnic cleansing, but because it was a UN-demarcated, NATO-enforced "safe area." Gorazde's significance increased after the initial NATO pinprick attacks on Serbian positions and the subsequent murder of a British UN soldier, attack on a French NATO aircraft, and downing of a British NATO Sea Harrier by Serbian forces. Gorazde became the site where the UN and NATO's credibility was at stake. The UN-NATO ultimatum effectively rezoned its meaning from a "safe area," a place where air strikes would be used to prevent attacks on UN operations, to a "safe haven," a place where military force would be used to shore up UN-NATO credibility in the region, although this distinction was not promoted by either the United Nations or the mass media.[86] Rather than targeting the practice of ethnic cleansing, the plan called for the targeting of the weapons and logistics of war. Nevertheless, the truly significant consequence of this was that the UN and NATO finally if reluctantly committed themselves to preventing attacks on civilians by raising the price of Serb shelling of the Muslim populations in "safe areas." As soon as the siege of Gorazde was lifted, however, UNPROFOR commander General Michael Rose publicly sought to discredit the testimony of those who articulated the horror of the siege, as part of his war of position against any moral imperative congealing in a way that would threaten the fiction of the moral neutrality of UNPROFOR's "humanitarian" mandate.[87]

Failing to Impose a Settlement: May–July 1994

The Gorazde debacle gave a new impetus to the efforts of the large powers to find some kind of a political solution to the Bosnian problem in the summer of 1994. Out of the diplomacy surrounding the crisis, a "Contact Group" composed of diplomats from the European Union, Russia, and the United States (represented by Ambassador Charles Redman) was institutionalized. In the course of negotiations with this group, the Clinton administration's policy changed in a number of ways. First, the United States moved closer to the European (largely French) position and, despite pronouncements never to impose a settlement upon the Sarajevo government or to get involved in drawing up a map, agreed to propose a 51–49 percent partition of the country, after having previously endorsed a 58–42 split.[88] In return, the French reportedly agreed to keep their peacekeeping forces in UNPROFOR for the immediate future. The U.S. position on the use of its forces in peacekeeping operations was codified in Presidential Decision Directive (PDD) 25, which was publicized by administration officials in early May. The directive was prompted by the Somalian "debacle" and attempted to specify strict conditions (American interests, availability of troops and funds, the necessity of U.S. participation, Congressional approval, clear exit strategy, and appropriate command and control arrangements) and precisely specified scenarios (international threat, gross abuse of human rights, and overthrow of a democratically elected government) that must be applied and met before the United States commits troops to international peacekeeping operations. The policy marked a significant move away from the "assertive multilateralism" (the proactive use of U.S. military and logistic experience in UN peacekeeping and peacemaking activities around the globe) outlined by candidate Clinton during the 1992 presidential election. Effectively, it codified a proactive unilateral isolationism on the part of the U.S. military in conflicts and regions, from Angola to Cambodia, Somalia, Rwanda, and Bosnia, where "vital U.S. interests" were not at stake.

Second, the United States changed its foreign policy staff. Steven Oxman, assistant secretary of state for European affairs and the person who coordinated U.S. policy on Bosnia, was dismissed and replaced by Richard Holbrooke. Jenonne Walker, a strong critic of the Bush administration policy on Yugoslavia and supporter of greater U.S. military involvement in Bosnia, was removed from her position as the European specialist on the National Security staff and became U.S. ambassador to the Czech Republic.[89] Walker was known to be critical of British and French policy attitudes toward Bosnia. David Gergen, the public relations specialist, was also shifted to the State Department, in a move that suggested that the Clinton administration's Principals saw their foreign policy difficulties partly, at least, in terms of public relations problems. Certainly, after PDD 25, there was an even greater awareness of the need to contain the impact of television images coming from such killing grounds as Bosnia and, most horrifically in the summer of 1994, from Rwanda. Clinton administration spokespersons, for example, were instructed not to use the term "genocide" in their foreign policy articulations on Rwanda lest it create an "inappropriate" moral imperative to act.[90]

Third, the Contact Group developed their own map for the partition of Bosnia and presented it to both the Muslim-Croat Confederation and the Bosnian Serbs with a deadline attached. Failure on the part of the Muslims to accept it would result in the withdrawal of UNPROFOR and the easing of sanctions on Serbia. Failure on the part of the Bosnian Serbs would result in a lifting of the arms embargo, an embargo that was already eroding as a result of the movement of arms into Bosnia through Croatia, against the Bosnian government. When the Muslims accepted the plan and the Bosnian Serbs rejected it, the Contact Group's divisions were exposed and nothing was done. Bosnia was not worth it.

GEO-POLITICAL SCRIPTS AND THE SOCIAL PRODUCTION OF MORAL (IN)VISIBILITY

The Bosnian war is an irreducibly modern war over space, territory, and identity. With its brutal and criminal campaigns of "ethnic cleansing," with Serbian forces pushing for lebensraum and an ethnically pure state while victims flee in terror and crowd into so-called safe areas or scatter into exile, the Bosnian war has given rise to geographical sites of purity and displacement, geographies of essential territorial identities, besieged enclaves, entrenched positions, bloody frontiers, and ubiquitous flows of refugees, humanitarian aid, medical supplies, and military weapons. The Serb drive to create an idealized, homogeneous state by expelling those they consider to be outsiders, Muslim infidels in Europe, is a fascistic form of modernity,[91] a deadly intolerance of ambivalence that, as Zygmunt Bauman has argued, is a practice inherent within the Western "gardening state."[92] Under the gardening gaze, the modern state is a territory that must be made to conform with a certain preconceived ideal or design, a design, in the case of Bosnian Serbs, for a homogeneous and ordered Serbian community secure in its own identity and space. Operationalizing such a vision requires acts of brutal intervention against those that despoil the vision or stubbornly resist its artificial dream of harmonious order. Such instrumentally rational acts are conceptualized as positive and creative, a separating, removing, and killing of those "human weeds" that blight the envisioned aesthetic ethnic garden. Physical territory is occupied, human habitats are cleansed, and symbolic landscapes are purged of any trace of the polluting Other, the outsider, the ethnically impure.

The ideal of the garden state is a geo-political vision that leads directly to acts of genocide. The reappearance of such a fascistic form of modernity and its genocidal project upon the European continent in 1991 has not, however, elicited an active military response from those powers who waged war against the earlier fascism of Nazi Germany. The reason for this can be found in the way they have chosen to write "Bosnia" as a place in their governmental maps. For the U.S. foreign policy establishment, "Bosnia" is, in the last instance, mapped as a potential quagmire. Ultimately the place is outside the universe of obligation of U.S. foreign policy. It is beyond the limits of duty, beyond the boundary within which open-ended, conditionless moral responsibility is exercised. In sum, "Bosnia" is not geopolitically close to the United States; it is a "faraway country."

The governmental mapping of "Bosnia" in such a manner is an act of geo-power that socially produces *moral invisibility*. The recognition of "Bosnia" as a site of genocide and a late-twentieth-century form of fascism has been consistently blocked by the instrumental reasoning of institutions such as the Pentagon, the State Department, and NATO. The bureaucratic culture of such institutions produces, as a matter of course, a moral indifference and a moral invisibility to the victims of ethnic cleansing in Bosnia. The expression of such moral indifference is found in the World War I/Vietnam/Somalia scripts, the reading of the contemporary Bosnian conflict as an ancient "blood feud" that is centuries old, and the writing of the region as an inherently murky, complex, and violent place, a morass, a tinderbox, a dangerous "womanplace" that does not allow Reason to take form, a location where one can have no moral compass.[93] What such forms of geographical reasoning do is geo-politically space "Bosnia" beyond the bounds of Western understanding and moral responsibility. They are scripts that socially produce distance. They are antihistorical forms of history and an antigeographical form of geography. They represent an intellectual failure of historical and geographical knowledge, a failure to see the open and undecidable character of regions and places.

It is not difficult to site "Bosnia" in terms other than those offered by the Pentagon or the decision-making elites in the Bush and Clinton administrations. The work of dissident diplomats and courageous reporters such as Misha Glenny, Roy Gutman, Maggie O'Kane, David Rieff, Ed Vulliamy, and numerous others does not produce an unmediated Bosnian actuality but a "Bosnia" understandable within the general history and practice of genocide in the twentieth century. The newpaper (and occasionally television) journalist Maggie O'Kane's "anti-geopolitical eye," in particular, offers an alternative way of seeing and narrating the Bosnian conflict. Rather than adopting the detached perspectivalism of diplomats, she brings a ground-level traveling eye to bear upon the landscape of the conflict. Her eye records the fractured lives and broken bodies of the victims of the war, lives that fall between the lines of official governmental cartographies of the war. She turns her own personal encounter with Bosnia and its people into a geopolitical journey, giving voice in her columns to those brutalized by ethnic cleansing and mass rape. In multiplying perspectives in this way, she destabilizes the governmental eye that sees Bosnia in terms of imperatives to dam off, bandage up, and block off its flowing feminine excesses, to contain and limit. O'Kane's anti-geopolitical eye breaches the governmental dam erected to separate and manage the "us here" from the "them there," thus letting Bosnia bleed into our world, onto our newspapers, and on our television screens. With the intense power of her prose, she corrodes the self-protective layers of indifference of the West. She writes not about an "over there" but about a global "here and now," a collapsing-boundary Bosnia bleeding into Britain and Europe's self-image. She reveals the moral quagmire within the West's own imagination of itself, within state and leaders that profess themselves moral yet send troops to be neutral in a country where genocide is occurring.[94] Bosnia, in other words, is made morally visible.

The scripting of "Bosnia" as a modern-day genocide is, of course, an incomplete and always questionable reading of the specificity of the conflict. It does nevertheless introduce points of unease into the assurance of governmental understandings of the conflict.[95] Television images of shelling and the grim consequences of ethnic cleansing are, likewise, not unmediated screenings of the Bosnian actuality but ambiguous and undecidable systems of signification. Yet they, too, have a power to potentially remap the place of Bosnia in the U.S. geo-political imagination. If the quagmire and associated scripts function to produce distance, the holocaust and associated scripts offer the possibility (although never the guarantee) of producing proximity, of reclaiming Bosnia within a universe of obligation. In the daily volume of sights and cites from Bosnia, we are offered the possibility of "Bosnia" (and ourselves) being made morally visible.

The ongoing tension between the social production of distance by quagmire scripts and the social production of proximity by holocaust scripts has already produced a peculiar geography of international and U.S. involvement in Bosnia so far. The United States is the key power underwriting the cost of the "humanitarian mission" in Bosnia (although its enthusiasm and willingness to pay wavers). It confines its military contributions to airpower, and it has occasionally been willing to use it to enforce the changing concept of "safe areas." U.S. fighter aircraft have bombed the Serbs, yet the United States has scrupulously sought to assert its neutrality in the conflict.

Whereas the social production of distance has conventionally been associated with heterophobia and the social production of proximity with responsibility, U.S. foreign policy practice has been paradoxically distinguished by a circumscribed "humanitarian" vision of ethical responsibility combined with an assertive social distancing. U.S. foreign policy has been characterized by responsibility and heterophobia. "Bosnia" is within a certain universe of moral obligation of U.S. foreign policy, but this moral obligation manifests itself in a technical and logistically defined responsibility and not in an unconditional moral responsibility. The United States provides funds for refugee assistance programs, airdrops food and medical supplies, funds medical equipment, and, maybe when the time is right, will provide troops to keep an already established cease-fire. It tries to use its military power to compel Serb forces not to attack Muslim civilians, but it does not and will not take an unconditional moral stance in the war. *It will not check ethnic cleansing because it is ethnic cleansing.*

Disclosed by this circumscribed geography of engagement is a nasty paradox at the heart of the concept of "humanitarianism." By siting "Bosnia" as a "humanitarian catastrophe" first and foremost, by focusing so much of its effort on the "humanitarian" dimension of the Bosnian war, U.S. foreign policy, together with that of the European powers and the United Nations, is acquiescing in an inhumane outcome: the victory of the policy of "ethnic cleansing."[96] The concept of humanitarianism, as articulated by these institutions, corrupts humanitarianism. Hiding behind their mandate, UNPROFOR and the West have become accomplices to genocide.[97] A moral approach to Bosnia can only be an unreasonable one, an antihumanitarian one that denounces UNPROFOR's humanitarianism because it is not sufficiently humanitarian.[98] In reflecting on the Jewish Holocaust, the distinguished German historian Hans Mommsen described its key anthropological dimension as "the danger inherent in present-day industrial society of a process of becoming accustomed to moral indifference in regard to actions not immediately related to one's own sphere of experience."[99] This eminently geographical lesson is, unfortunately, particularly apt when considering U.S. foreign policy toward the dilemma that is "Bosnia." Rather than seeing "centuries-old" ethnic hatreds *in* our symbolically constructed "Bosnia," perhaps we should consider the view *from* Bosnia and what it shows us about our own indifferent, postindustrial, and postmoral society.

7

Visions and Vertigo
Postmodernity and the Writing of Global Space

> The fundamental political conflict in the opening decades of the new century, we believe, will not be between nations or even between trading blocs but between the forces of globalization and the territorially based forces of local survival seeking to preserve and to redefine community. — RICHARD BARNET AND JOHN CAVANAGH[1]

GEOPOLITICAL VERTIGO

The end of the Cold War has caused a crisis of meaning in global politics. The collapse of Communism and the demise of the Cold War has deprived international affairs of an organizing script and a defining drama. Many other scripts of international affairs are apparent— the destruction of the planetary environment, the resurgence of xenophobic nationalism, the rise of religious fundamentalism—but no one script predominates. The issue is not that global politics lacks meaning. "The problem," according to one figure, "is not that there are no reasonable contending definitions of the new era but rather that there are too many of them."[2] For some, as we noted in the last chapter, Bosnia is the quintessential site of the post–Cold War; for others, it is the collapse of the state in Africa, the Algerian civil war, terrorism in Oklahoma city, or the prospect of global warming.

For a foreign policy establishment comfortable with a Cold War world it helped write, the end of the Cold War world inevitably induced a certain experience of disorientation and vertigo.[3] Events accelerated with remarkable speed. The old rules of interstate politics no longer applied. Things seem to be out of control. Zbigniew Brze-zinski, in a book entitled *Out of Control: Global Turmoil on the Eve of the Twenty-First Century,* articulates this structure of feeling in writing of "the notable acceleration in the velocity of our history and the uncertainty of its trajectory." "History has not ended," Brze-zinski observes, "but has become compressed. Whereas in the past, historical epochs stood out in relatively sharp relief, and one could thus have a defined sense of historical progression, history today entails sharp discontinuities that collide with each other, condense our sense of perspective, and confuse our historical perceptions."[4] History, for the foreign policy establishment at least, is no longer a fixed representational scene but a vertigo-inducing experience.

This condition of *geopolitical vertigo* — as the global affairs magazine *New Perspectives Quarterly* once described it[5] — is in keeping with a more generalized structure of feeling associated with the condition of postmodernity. This condition, according to David Harvey, is a historical-geographical condition bound up with the emergence of newly dominant ways in which we experience space and time, an emergence Harvey connects with a worldwide transition from increasingly redundant national Fordist structures of production to new flexible post-Fordist modes of capital accumulation on a global scale.[6] Whether we can explain the condition of postmodernity in quite the structural manner Harvey does or not, the new round of time-space compression he identifies (an intensification of an already existing vertigo of the modern) has promoted a widespread cultural and intellectual structure of feeling characterized by disorientation and dislocation. Newly proliferating forms of communication and media have not only intensified the capitalist culture of reification and distraction but have also induced information saturation among its governing elites. Everyday life seems to be irradiated with too much meaning, with an abundance of mediated realities. The intensity and speed of the political, economic, and cultural transformations wrought by a globally organized, flexible accumulation threaten to overwhelm the ability of intellectuals to assimilate them.

Within the Western foreign policy community, this experience of disorientation and vertigo has important implications for the possibility of geopolitics after the Cold War. As a form of Cartesian perspectivalism, geopolitics enframed the world within modern Western categories and conceptions. Geopolitics produced international politics as theater: geography was the stage, politics the drama, and geopolitics the detached observation of this representational spectacle. As the positioning that made this project possible is destabilized in postmodernity, the geopolitical production of global space as a modern perspectivalist scene/seen becomes increasingly problematic. The very perspectivalist notion of a *scene* of global politics — one that relies on a notion of detached observation by a Cartesian subject separated from an objective external "seen" — seems imperiled. Postmodernist theorists have long argued that living in a society of generalized communication, a society of mass media, has certain consequences for how we conceive and perceive reality. For Gianni Vatimo, the rise of telematic communication — newspapers, radio, television, and the worldwide webs of the Internet — has been decisive in bringing about the dissolution of centralized perspectives, what Lyotard calls "grand narratives."[7] He suggests that "the increase in possible information on the myriad forms of reality makes it increasingly difficult to conceive of a *single* reality" (7). With the proliferation of images, we are losing our bearings and sense of reality. According to Mackenzie Wark, we no longer have roots, we have aerials; we no longer have origins, we have terminals.[8] In his typically hyperbolic manner, Jean Baudrillard proclaims that the order of representation has given way to the order of simulation; the theatrical scene has been supplanted by the television screen. Reality no longer has any depth. "There is no longer transcendence or depth, but only the immanent surface of operations unfolding, the smooth and functional surface of communication."[9] With the erosion of the distinction between the subject and the object, the interior and the exterior, the near and the far, reality becomes *obscene*, a forced, exaggerated, close-up pornography of images. In a passage that anticipates the microscopic videography of such mediated events as the Gulf War, the Goma refugee crisis, or the O. J. Simpson trial, he suggests that obscenity "begins when there is no more spectacle, no more stage, no more theater, no more illusion, when everything becomes immediately transparent, visible, exposed in the raw and inexorable light of information and communication" (21–22; original emphasis). The obscene is not that which is hidden but the all-too-visible, the more-visible-than-visible universe of communication. Detached distance has been replaced by an overproximity to giddy images in perpetual orbital transmission. The mise en scène of classical geopolitics appears to have given way to the *mise en abŷme* of simulated geopolitics, the hyperreal environments of geographical information systems, war game simulations, global television emissions, and cinematic sign systems, all of which place meaning in limbo.[10]

Jean Baudrillard, however, is an imperfect guide to postmodernity. His work is, as Douglas Kellner suggests, best read as science fiction rather than social theory.[11] Provocative, hyperbolic, and apocalyptic, Baudrillard exaggerates the break between the modern and the postmodern (the order of representation and the order of simulation) and writes — with a nostalgia for a lost authenticity — about the sign economies of flexible capitalism from within. As a consequence, his work ignores contemporary capitalism's structures of power, class, and political economy, which create and perpetuate these sign economies. His message is no more than the media he comments upon: sound-bite theory that is sometimes interesting and entertaining but often outlandish and absurd.[12] His rejection of the possibility of critique of the status quo leads him to a mindless politics of fatalism. The experience of vertigo, for example, is an absolute condition and not an analyzable conjuncture, a position that is effectively anti-intellectual in that it promotes intellectual surrender in the face of the confusion and flux of the present.[13]

Postmodernity, however, is a materialist, historical-geographical condition that can be understood with critical reason. Baudrillard's use-value is in recognizing and elucidating how the meaning of space and territory are increasingly in crisis within this condition. While Baudrillard's observation that maps (signifiers) now precede our knowledge of territory (the signified)[14] is hardly new given the history of the modern state system, his unique presentation of the rarefaction of reality — the intensified delinking of signs from their ostensible referents — is worth taking seriously. This is particularly important in considering space and territory, for what happens to both is crucial to the problematic of postmodernity. The emergence of a condition of postmodernity over the last three decades has coincided with a dramatic materialist and ideological *deterritorialization* of the geopolitical world order established under American hegemony after World War II.[15] This process of deterritorialization refers not to the creation of a "borderless world" but to the loosening of the spatial order (forms, hierarchies, and codes) of Pax Americana from the late 1960s onward. This loosening and scrambling of the postwar spatial order is a result of both the diminishing power of territorial states in world affairs generally and the relative decline in the power and ability of the U.S. state to master global political space in particular. This deterritorialization is most evident in finance and production. First, the collapse of the Bretton Woods system of fixed exchange rates, the growth of offshore banking centers, and the tremendous expansion of xenocurrency markets (the Eurodollar markets and others) created a global financial system beyond the mastery of any hegemon or concert of states.[16] The failure of statist regulatory efforts to keep pace with technological changes and market innovations (the creation of new forms of credit and debt) in the 1980s produced a "casino capitalism" characterized by ever-increasing velocities of capital flows, daily turnover rates, and currency market speculations.[17] Although the U.S. dollar remains the world's leading reserve currency, the ability of the central banks of the Group of Seven states to control private international financial markets is now seriously in doubt. Second, the move from a Fordist to a post-Fordist regime of accumulation has been accompanied by a gradual dissolution of national economic space and an intensifying globalization of manufacturing and services. According to Robert Reich, there is no longer any such thing as unitary national economies. Globalization is undoing the connection between national core corporations and the territorial economies of states. Former national champions (like Volkswagon, General Motors, British Petroleum, or Phillips) are no longer automatically loyal to their country of origin. Such corporations may still be nominally national, but they are functionally transnational. Their health is a consequence of the strength and profitability of their global webs.[18] As a result of this globalization, the previous solidity and coherence of territorial economies are beginning to come apart and bifurcate into zones connected to global webs and flows ("symbolic analyst zones," according to Reich) and zones outside and disconnected from legal global webs and flows (although not from the illegal webs of the global drug industry).

While it is important not to exaggerate the degree of globalization and deterritorialization, these and other material transformations have rendered the rigidities of the modern sociospatial triad of the interstate system (state sovereignty, territorial integrity, and community identity) increasingly problematic. This heavily mythologized triad of state-territory-community was never perfectly set and stable in any country, but its instability and precariousness are becoming more and more pronounced as places are denationalized and globalized by transnational flows. These tendencies were occurring well before the dramatic collapse of Communism and subsequent disintegration of the territorial organization of the Eastern bloc and Soviet empire after 1989. This implosion of the geopolitical order of the Cold War starkly foregrounded the degree to which the post–World War II world order had come apart and placed the meaning of the "West," "Europe," and the "United States" as sociospatial identities in crisis, thus provoking the experience of vertigo we have noted. But every deterritorialization creates the conditions for a reterritorialization of order using fragments of the beliefs, customs, practices, and narratives of the old splintered world order. Out of the experience of vertigo, newly imagined visions of state, territory, and community are projected in an effort to restabilize and reterritorialize identity amid global flux. As one order of space unravels, new orders are deployed to retriangulate local foregrounds against global backgrounds into new productions of global space.

AMERICAN VERTIGO

This chapter examines two neoconservative attempts to reterritorialize the identity of the United States and rewrite the meaning of the global in response to the deterritorialization of the Cold War and the vertigo of the condition of postmodernity.[19] The two figures I wish to consider—Edward Luttwak and Samuel Huntington—are both neoconservative Cold War intellectuals who strive to step back from the current confusion and dizziness of the post–Cold War scene and consciously produce a vision of the crucial geo-lines of conflict in global politics at the end of the twentieth century. In the writings of both we find a common anxiety about the geopolitical vertigo and spatial blurring of the condition of postmodernity. Both intellectuals are uncomfortable with the passing of the familiar world of the Cold War, a geo-political congealment that was more imaginary than real yet nevertheless legitimated the creation and maintenance of a society of security safeguarding the "United States" and the "West" in the postwar era.[20] This society of security—a society of which Luttwak and Huntington are a part—is endangered by a deterritorializing world order where clear and present dangers are not manifest and obvious. In response to this, both intellectuals reterritorialize global space with threats to the security of an economically enfeebled and culturally divided United States. In Luttwak's case, the main threat is an economic one, and it comes from a capital-rich and technologically superior Japan and, to a lesser extent, from a determined, subsidizing Europe. In Huntington's case, the threat comes from the "non-West," specifically a "Confucian-Islamic" bloc of states (although he too is concerned with the economic threat posed by Japan). In both cases, there is also considerable emphasis on the need for domestic economic and cultural renewal within the United States. Despite considering globalization and in keeping with postwar strategic culture, both intellectuals write global politics as a zero-sum struggle for power between competing states or state-like entities. In examining both writings of global space as acts of geo-power, I wish to argue that both produce a global political scene following the lines and triangulations of past Cold War calibrations of danger. They write "new" national security jeremiads with old script lines in an effort to revitalize the society of security and perpetuate the culture of the cold war despite the end of the Cold War as a historical period.[21] The Cold War may be dead, but cold war strategic culture and its society of security is undead. A vampirish culture, it lives on by sucking meaning out of the new global indeterminacies and contingencies as well as material resources for itself from the depleted reserves of society as a whole. After examining Luttwak and Huntington's productions in detail, I conclude by briefly considering some of the implications of the crisis of the sociospatial triad of the interstate system for the future of global politics into the twenty-first century.

EDWARD LUTTWAK: THE VERTIGO OF
THE THIRD WORLDIZATION OF AMERICA

Dr. Edward Luttwak is a quintessential "defense intellectual." A native of the Hungarian region of Transylvania, he is director of geoeconomics and senior fellow at the Center for Strategic and International Studies in Washington, D.C., as well as an international associate at the Institute of Fiscal and Monetary Policy of Japan's Ministry of Finance. Since the 1970s Luttwak has made a career for himself as a "defense expert" and consultant to various American and European corporations. In the early seventies he was associated with the Tevel Institute in Israel, an institute later revealed to be funded in part by the U.S. Central Intelligence Agency.[22] Once in the United States, Luttwak became part of a pro-Israel, antidétente community of Jewish neoconservative intellectuals in Washington, D.C., who campaigned against SALT and in favor of enlarged defense budgets. Luttwak's early books included *The Grand Strategy of the Roman Empire* (1976) and *The Grand Strategy of the Soviet Union* (1983), both of which treat "grand strategy" as a transhistorical practice with a timeless essence. His *The Pentagon and the Art of War* (1985) is a critique of Pentagon bureaucracy and force structure, although not of the defense contractors who paid his consultancy fees.[23]

With the end of the Cold War, Luttwak discovered "geo-economics" and retooled and remarketed himself as a new geo-economic strategist who could expertly read the armory, battle formations, and strategic stakes of an emergent worldwide struggle for industrial supremacy among states. Luttwak first outlined his understanding of geo-economics in an article in *National Interest* in which he specified geo-economics as the continuation of "the logic of conflict in the grammar of commerce." Contrasting geopolitics to geo-economics, he asserted that the methods of commerce are displacing military methods in international relations, with disposable capital becoming more important than firepower, civilian innovation more significant than military-technical advancement, and market penetration a greater mark of power than the possession of garrisons and bases.[24] He subsequently elaborated this thesis in the book *The Endangered American Dream* (1993).[25] Subtitled "How to Stop the United States from Becoming a Third World Country and How to Win the Geo-Economic Struggle for Industrial Supremacy," Luttwak's book argues that "old-fashioned" geopolitics has been displaced by the new phenomenon of geo-economics. Guns and diplomats now count for much less than patient capital and highly skilled labor. A different arsenal of weapons characterizes this new version of the ancient rivalry of states: aggressive national technology programs, predatory finance, and ambushlike tariffs and technical standards. Unless the United States takes dramatic action to bolster its economic power, it will suffer defeat after defeat and could ultimately drop to the status of a Third World country. This apocalyptic vision is typical of the alarmism found in certain geo-economic literatures from the late 1980s.[26] Like much of this literature generally, he uses a jeremiad mode of representation to organize the textuality of global economic transformations into a crisis in need of response. The United States is fighting a permanent war against a determined enemy. Some are not even aware of this global struggle, yet the American way of life is endangered by this conflict. America needs to wake up and rediscipline its slackening, errant, and divided self. Otherwise disaster will befall all.

Luttwak's account begins not with a panoramic survey of space but with the experience of a spatial journey by an imaginary male traveler from Tokyo to New York. (The imaginary traveler is clearly Luttwak himself—he holds positions in both the United States and Japan—but the device allows him to write his own personal experience as the norm.) After leaving the taxi at Tokyo's downtown City Air Terminal, which, Luttwak remarks, is "a perfectly ordinary Tokyo taxi and therefore shiny clean, in perfect condition, its nearly dressed driver in white gloves," he will find himself on "an equally spotless airport bus in five minutes flat"; on the bus he is professionally and efficiently processed for boarding onto the plane, which departs following schedules timed to the exact minute. From the ultramodernity of Japan, the traveler reaches JFK in New York and is "confronted by sights and sounds that would not be out of place in Lagos or Bombay." Luttwak continues:

> Instead of the spotless elegance of Narita or Frankfurt or Amsterdam or Singapore, arriving travelers at one of the several terminals that belong to near-bankrupt airlines will find themselves walking down dingy corridors in need of paint, over frayed carpets, often struggling up and down narrow stairways alongside out-of-order escalators. These are JFK's substitutes for the constantly updated facilities of First World airports. The rough, cheap remodeling of sadly outdated buildings with naked plywood and unfinished gypsum board proclaim the shortage of money to build with, of *patient* capital avail-

able for long-term investment—although there was plenty of money
for "leveraged buy-outs" and other quick deals in New York during
the years that JFK decayed into a Third World airport. (15–16)

The passage is a remarkable snapshot of spatial disorientation and
vertigo. The East's modernity surpasses that of the West. Japan is
what the West images or desires itself to be: clean, ordered, and efficient in its management of time. The nominal West (New York) has
become a chaotic Orient, a place where the white male body is as-
saulted by the sights and sounds of dilapidation. The frayed carpets,
the defective elevators, and the pervasive dirt are "the perfect sign
of Third World conditions, the instantly recognizable background
of South Asian, African, and Latin American street scenes" (16). In
contrast to Tokyo, JFK is characterized by inexcusable delays and
an inefficient management of time. Porters unashamedly solicit tips.
In New York, the taxis are "usually driven by an unkempt, loutish
driver who resembles his counterparts in Islamabad or Kinshasa
rather than in London or Tokyo, where licensing requirements are
strict and dress codes are enforced" (17). The ordered spatial hierarchies of the Cold War have given way to a condition of deterritorialization, a spatial blurring of the First and Third Worlds.

Luttwak's observations are prompted by the dislocation of high-
speed air travel and the interchangeability of places in the condition
of postmodernity. It is an airport, a postmodern space of flow par
excellence, that triggers his spatial anxiety attack. The Tokyo–New
York air corridor is made into a symbolic journey from a modern
Japan to a confusing, disassembling "United States." This route is
no longer a unilinear journey from a traditional "there" to a modern "here." The "United States" is now a postmodern space, meaning a less modern space to Luttwak. It is a place out of sync; its previously unilinear progress is suspended, fragmented, eclipsed by a
chaotic "now." Luttwak begins his imaginary journey from Tokyo
but suggests it would be much the same if he began in Zurich, Am-
sterdam, or Singapore. These foreign airports are imaginary semiotic
spaces, spaces outside a dissolving and blurring United States. The
sign "Japan" functions as the point of departure of a new geo-
economic myth-history; it is an idealized, hyperreal space, an imaged
semiotic Japan (or, occasionally, Europe) produced by growth stories,
economic statistics, and success narratives that smooth out its own
economic shortcomings and spatial incoherences. With an imaginary

Japan as a semiotic anchor, Luttwak begins to construct a hyperreal America that is under threat from its failure to manage globalization. Dilapidated corridors in JFK are forced to work as symbols of a deep malaise within the United States. The "naked plywood and unfinished gypsum board at Kennedy airport are perfect symbols" of an American capitalism that is naked, a capitalism without capital (118).

To most experienced international travelers, Luttwak's comments about JFK are recognizably exaggerated. A single terminal or corridor under reconstruction and repair is forced into service as a symbol of the structural condition of an "American capitalism" that is still assumed to be a recognizable coherency by Luttwak. The passage through JFK is symbolic of America's own passage into the contingencies of the post–Cold War age, a transition marked by a sense of loss at the passing of the old certainties of the Cold War. America itself is in need of repair. Its previously well finished corridors of power are now fraying.[27] Japan is the new myth around which "America" must reconsolidate itself as a solid presence, not a dissolving, fading, and fragmentary coherence.

The horizon of Luttwak's vision of the United States at the fin de siècle is his claim that global politics has now entered a geo-economic age. The dynamic generating this horizon of geo-economics is the timeless, permanent struggle of states for power. Geo-economics is

> not more and not less than the continuation of the ancient rivalry of nations by new industrial means. Just as in the past when young men were put in uniform to be marched off in pursuit of schemas of territorial conquest, today taxpayers are persuaded to subsidize schemes of industrial conquest. Instead of fighting each other, France, Germany and Britain now collaborate to fund Airbus Industrie's offensive against Boeing and McDonnell-Douglas. Instead of measuring progress by how far the fighting front has advanced on the map, it is worldwide market shares for the targeted products that are the goal. (34)

Feelings of hostility and animosity between nations are also written as a permanent feature of international politics by Luttwak. However, in the post–Cold War world, the form of expression of these feelings now has a distinctive geography. In marginal places, people still fight old-fashioned military wars, whereas in the core places of the world, people fight economic wars. "In the backwaters of world politics, where territorial conflicts continue, wars or threats of war provide an ample outlet for hostile sentiments. But when it comes to the central arena of world politics where Americans, Europeans, and Japanese collaborate and contend, it is chiefly by economic means that adversarial attitudes can now be expressed" (35). Tax policies, subsidies to industry, hidden tariffs, quotas, voluntary restraint agreements, and regulatory requirements are all instruments in this geo-economic battle between the best in the West, a battle that is possible because all the participants have already ruled out the possibility of old-fashioned military war among themselves.

This vision of an advanced world characterized by geo-economic conflict leads Luttwak to take readings of the position of the United States vis-à-vis its competitors. In contrast to Japan, the process of globalization has not been managed well by the U.S. government. Globalization has not been controlled and offset in the interests of the vast majority of Americans. Rather, too often globalization is being managed to "benefit corporate management, shareholders, and elite design and development employees." Japan, by contrast, has managed globalization to secure production jobs for Japanese workers, manufacturing earnings for its corporations, and technological leadership at the expense of America (21). The differential benefits of globalization within the United States has led to a situation where the income levels of "non-elite Americans and their families" are sinking to an ignominious parity with their Mexican counterparts. After reviewing the relative decline in U.S. wages and incomes since the 1970s, Luttwak concludes that the United States may remain a First World country, but a distinct majority of all Americans "have long been headed 'south' in stock-market parlance" (124). A growing U.S. underclass already lives in Third World conditions, while a small elite of Americans act like a Third World–style rentier class and live off global stocks and bonds investments. They are unwilling to pay for public investment with their taxes and contribute little to the productivity of the economy.

Luttwak finds much to fault in how the United States is organized. First, opinion leaders and policymakers in the United States are held in the grip of a free trade ideology that is invalid in the practical political world of global economic competition. Free trade confuses the American mind and inhibits necessary action. The result is that Japanese corporations are able to take advantage of the open nature of the U.S. economy, while U.S. corporations are largely shut out from operating in Japan. Second, U.S. deficit spending and conspicuous consumption have resulted in a lack of patient investment capital in the United States available for long-term strategic investment projects and tasks. U.S. industry often fails to reinvest its profits productively, or it reinvests its profits in building manufacturing plants overseas. The Calvinistic spirit of previous U.S. generations has disappeared, a condition Luttwak medicalizes. American society is stricken by a dreaded ACDS (Acquired Calvinist Deficiency Syndrome) and a pervasive shortage of capital, the chief source of advancement of any society, ancient or modern (248). Third, the U.S. educational and legal systems are grotesque failures. Besides lacking the national educational standards of a state like Japan, the U.S. educational system is hindered by a multiculturalism that is racist, absurd, and countereducational (271–87). American culture, with its laws and its norms, has also "decisively tilted towards an unrestrained individualism that knows no balance" (214). Individual wish-fulfillment is pursued at the expense of family and community. Government agencies are guilty of legalistic extremism, while private lawyers and product liability laws erode the competitiveness and innovative potential of U.S. industry. While Luttwak does suggest a series of what he considers solutions to these problems, his overwhelming concern is with the description of U.S. weakness. Following the presentation of a lengthy set of statistical comparisons between the United States and competitive geo-economic states, principally Japan, Luttwak turns to a second allegorical vision:

Having until now relied on abstract statistics, to escape the influence of mere impressions too often unrepresentative, it is interesting simply to look at the summer-time street scene just outside the building where these lines are being written. It is excessively symbolic that alongside the "postindustrial" office buildings and luxury hotels of Connecticut Avenue and K Street NW—the central downtown intersection of Washington, DC, the nation's capital and arguably the world's—one may see on any working day: (1) a white, blonde mother-with-baby sitting on the pavement to beg in full Calcutta style; (2) the stands of licensed vendors of cheap clothing and African "airport art" that narrow the path left to the passing lunchtime crowds, just as in Istanbul until its recent street regulations were imposed; (3) several occasionally insistent panhandlers, some very colorfully dressed; and (4) a knot of unkempt vagrants who actually live in the Farragut North Metro entrance, which they occasionally befoul in full view of passersby, unless feebly interrupted by hapless transit police officers under court orders to let the "public" (i.e. vagrants) use public facilities "freely." (269–70)

Once again, we see expressions of spatial blurring and disorientation, concern with the breakdown of traditional grand narratives (law, order, and the family), and a preoccupation with people who are dirty and unkempt and who shamelessly solicit money. Like JFK, the U.S. capital is another symbolic space that is meant to embody and monumentalize the power of the United States but is blighted by a deep economic malaise that has social, political, and cultural dimensions. Both the JFK and Washington vignettes are not classic geopolitical displays of space that exhibit the world in a static panoramic fashion. Rather they are allegorical snapshots of the fraying and decay of an imperial map. They function as visual shock scenes that assault and offend the faculties and sensibilities of the (white male) strategist who remembers when things were different. They are "ob-scenes," scenes without imperial order and perspective, decapitalized scenes.

Ostensibly, of course, Luttwak's geo-economic narrative attempts to give coherence to international politics in the post–Cold War age, a politics that is still conceptualized as interstate, not global politics, a politics comprehensible within the logic of political realism. But Luttwak's geo-economic narrative is merely a projection of the myth-history of "Japan" upon world affairs without regard for the specificity and differential intensity of transnationalization across the globe. Luttwak's own descriptions of the fragmentation of the U.S. territorial economy and the disappearance of a consensual American national community contradict his overly confident assumption that "states will tend to act geo-economically simply because of what they are: territorially defined entities designed precisely to outdo each other on the world scene." What Luttwak describes but does not grasp is that space is no longer primarily mastered by nation-states. The territoriality of global affairs is no longer one of competing, segmented, and discretely sovereign nation-states but a territoriality shaped by global flows. These flows have produced a highly differentiated enclavization of previously homogeneous national territorial space, an enclavization Luttwak endeavors to describe in his discussion of Third World inner cities and First World gated communities of plenty within the United States. Yet, rather than extend this analysis worldwide and outline how global transnational flows are creating zones of wealth and privilege throughout the world, on one hand, and zones of poverty and endemic crime, on the other hand, Luttwak remains within the nation-state territoriality assumed by his ahistorical political realism.

Thus, while American corporations may be moving overseas and forming strategic alliances with other foreign corporations, he nevertheless speaks about "American" corporations, "American" industry, and "American" technology. Despite the evident deterritoriality of "American capitalism," he still assumes the fictional unities of an "American economy" and "American national interests." Luttwak, in other words, continues to strategize in statist terms despite his account of the increasingly rarefied nature of this space. One overwhelming reason why Luttwak still does so is his assumption that state bureaucracies are embracing geo-economics as a substitute for their military and diplomatic roles of the past. Geo-economics offers state bureaucrats a means of relegitimating themselves and their authority over their citizens (313). Thus, despite the evident globalization of previously unambiguous national champions and the dissolution of a unitary national community space, state institutions and bureaucracies will ensure that the territoriality of world affairs is an international and interstatist one and not a global, transnational one.

The problem with this argument, however, is that it underestimates the degree to which the American state has itself become transnationalized over the past two decades. The state as an entity is ahistorically conceptualized as a unitary institution that occupies and serves the interests of a particular territory. Yet Luttwak's own observations can be taken as evidence that this has not been the case in the United States in the last twenty years. Globalization, by his own admission, has not served the interests of the majority of people in the United States. To a considerable extent, the U.S. state under Ronald Reagan and George Bush delinked itself from its own territory by following a transnational liberal ideology and actively promoting transnationalization on a global scale. Federal government bureaucrats and officials in Washington, D.C., for example, are more interested in promoting the interests of nominally national but functionally transnational corporations in China, Vietnam, or Russia than in promoting the interests of certain tracts of the capital city itself, a capital city that has become a major distribution point on the global webs of the international drug industry and, as a consequence, the murder capital of the United States. To the extent that we can talk about a unified state in the United States, it has over the past two decades tended to serve the interests of mobile and footloose transnational capital and not the interests of territorially based citizen and neighborhood community groups. The U.S. state, in other words, has long been practicing a form of geo-economics, but it is a geo-economics that is promoting, not hindering, the deterritorialization and spatial blurring that Luttwak laments. Luttwak himself is part of that which he decries. He is not a national intellectual but a transnational symbolic analyst on the move.

The coherence of Luttwak's geo-economic narrative as an explanation of the world scene in the late twentieth century is thus very much in question. Luttwak's book is significant not for the explanatory power of his narrative but for its expression of the anxiety of the suddenly redundant neoconservative strategy caste in a world that has lost its dimensionality and shape. (It is their American dream that is endangered). As such, it is a record of a significant structure of feeling within an important strata of the U.S. foreign policy establishment, a feeling of loss and resentment at a world that has become uncomfortably deterritorialized and is in urgent need of the disciplinarity of a new geo-economic cold war.

SAMUEL HUNTINGTON'S CIVILIZING OF GLOBAL SPACE

In contrast to Edward Luttwak, Samuel Huntington is an intellectual of statecraft who specializes in questions of governance, particularly the problems of hegemonic governance. Educated at Yale (B.A., 1946), the University of Chicago (M.A., 1948), and Harvard (Ph.D., 1951), Huntington has spent his life within the elite circles of the Ivy League academic and foreign policy establishment. A neoconservative adviser to Democrats like Hubert Humphrey and Jimmy Carter, Huntington was retained by Harvard after his doctoral studies to work in its Department of Government. Cofounder of the journal *Foreign Policy,* associated at various times with Columbia, Oxford, and Stanford universities, a member of the Council on Foreign Relations and the Trilateral Commission, Huntington has served as a consultant to various agencies of the U.S. government and worked as coordinator of security planning at the National Security Council (1977–78) when that agency was headed by his one-time colleague and coauthor Zbigniew Brzezinski. Huntington was long associated with the Center for International Affairs (CFIA) at Harvard, a research institute established in 1957 by Robert Bowie (chief policy planner in the State Department of John Foster Dulles) together with McGeorge Bundy and Henry Kissinger.[28] Through its research fellowships, the CFIA was designed to attract and cultivate an array of influential intellectuals from around the globe who subsequently would serve as a worldwide network of informal influence for Harvard and its ideas of good government. Huntington served as associate director (1973–78) and subsequently as director (1978–89) of the CFIA. He is currently the director of the John Olin Institute of Strategic Studies at Harvard.

Huntington's work is explicitly concerned with questions of governmentality in developing and developed states. In *Political Order in Changing Societies* (1968), he addresses the question of how viable political regimes can be established and made to last in developing countries, a problem that was of particular concern to the U.S. government in South Vietnam at the time. Huntington's stress on the establishment of democratic organizations and institutions of authority (particularly the military) was a direct response to the challenge Communism posed to "modernizing countries."[29] Organization, he concluded, was the road to political power. "In the modernizing world he controls the future who organizes its politics" (461). In *The Crisis of Democracy* (1975), a report on the governability of developed democracies to the Trilateral Commission (at the time under the directorship of Zbigniew Brzezinski) in the wake of the sixties upheavals and Watergate, Huntington analyzes the social and political upheavals of the period in terms of a growing disrespect for authority and a general "excess of democracy." The solution to the adversarial culture of the youth, the mass media, and dissident intellectuals was "a greater degree of moderation in democracy" and restoration of a "democratic balance," code phrases for a neoconservative redisciplining of society around elitist and hierarchial notions of "expertise, seniority and experience."[30] Like other neoconservatives, Huntington was critical of his former colleague Henry Kissinger's policy of détente with the Soviet Union. He supported the

U.S. military buildup begun under President Carter and continued under President Reagan, as well as the U.S. war against the Sandinistas and its tilt toward the governments of El Salvador and South Africa.[31]

The sudden end of the Cold War in 1990 threw the world of the intellectuals and apologists of Cold War militarism into confusion. "The world changed in 1990," Huntington remarked, "and so did strategic discourse."[32] But, in Huntington's case at least, strategic discourse did not change that much. The Cold War with the Soviets may have ended, but there were suddenly new (and modified old) cold wars to be fought in the murky and dangerous post–Cold War world. This emerging world, Huntington noted, "is likely to lack the clarity and stability of the Cold War and to be a more jungle-like world of multiple dangers, hidden traps, unpleasant surprises and moral ambiguities" (7). Huntington outlines three principal American strategic interests: (1) maintaining the United States as the premier global power, which "means countering the Japanese economic challenge"; (2) preventing the "emergence of a political-military hegemonic power in Eurasia"; and (3) protecting "concrete American interests in the Persian Gulf and Middle America" (8).

Like Luttwak, Huntington securitizes the U.S.-Japan relationship and highlights the United States' "economic performance gaps" with Japan, evoking the specter of The Japan That Can Say No as evidence that Japan is an emergent threat to U.S. primacy in world affairs.[33] This focus on Japan as a threat and preoccupation with maintaining the primacy of the United States in world affairs led Huntington to support Bill Clinton for the presidency in 1992. For Huntington, the economic renewal of America was an overriding priority, but it was an economic renewal that was to be achieved not by downsizing the U.S. military or breaking up the society of security. Rather, the United States had to intensify its concern with security by improving its "competitiveness" and confronting Japan. For "the first time in two hundred years," Huntington wrote in a symposium on advice for a Democratic president, "the United States faces a major economic threat." "In terms of economic power... Japan is rapidly overtaking the United States. And economic power is not only central to the relations among the major states, it is also the underpinning of virtually every other form of power."[34] Japanese strategy is a strategy of economic warfare. Buttressed by appropriate citations by Japanese figures declaring themselves the new economic super-power and the United States as a premier agrarian power ("a giant version of Denmark"), Huntington reasoned that Chamberlain and Daladier did not take Hitler seriously in the 1930s, nor did Truman and his successors take it seriously when Stalin and Khrushchev said, "We will bury you," but Americans "would do well to take equally seriously both Japanese declarations of their goal of achieving economic dominance and the strategy they are pursuing to achieve that goal."[35]

Huntington's rewriting of the threats faced by the United States from Japan, Eurasia, and the Third World crystallized into a geopolitical world picture that was prominently unveiled and publicized in the Council on Foreign Relations journal *Foreign Affairs* in 1993. Entitled "The Clash of Civilizations," Huntington's essay was promoted—first as a leading article with solicited comments and reply in *Foreign Affairs*, second as a *New York Times* opinion editorial piece and subsequent syndicated columnist debate, and third as a special Council on Foreign Relations reader—as a comprehensive vision of the "next pattern of conflict" in world politics.[36] His goal, in his own words, was to produce "the best simple map of the post–Cold War world."[37] The overarching ambition, conciseness, and sloganistic simplicity of Huntington's mediagenic thesis accounts for its appeal to opinion makers, news journalists, and professional politicians casting about for a new interpretative system by which to order global affairs, given the waning interpretative power of the more optimistic visions of Francis Fukuyama and Bush's New World Order. In 1993, at least, Huntington's thesis was good copy with the media's foreign policy scribes.

Although it is provoked by the vertigo of postmodernity, Huntington's vision is projected with modernist assurance. The heteroglossia of global politics—what Huntington refers to as "bloomin' buzzin' confusion"[38]—is reduced to a total world picture. The gaze Huntington employs is that of a natural scientist qua geologist observing a world of solid forms whose meaning can be declaratively and unambiguously stated. He writes as a self-certain subject who reveals the "basic" plate tectonics of civilizational blocs (the "product of centuries") that "stretch back into history" and are now clashing once more along ancient "fault lines." Like a natural scientist, Huntington employs a series of definitional, periodizing, classificatory, and spatialization strategies to enframe human history and global space into an ordered geological exhibit. Civilizations, for Huntington, are foundational cultural totalities. A civilization is "the highest cultural grouping of people and the broadest level of cultural identity people have short of that which distinguishes humans from other species." Although they are supposedly centuries old and primordial, the clash of civilizations is the very latest stage in the evolution of conflict in the modern world, following monarchical, popular, and ideological stages of conflict. This fourth stage features a clash between seven or eight major civilizations, classified as Western, Confucian, Japanese, Islamic, Hindu, Slavic-Orthodox, Latin American, and possibly African. The clash of these civilizations occurs at a micro and a macro level, the microlevel being the struggle of adjacent groups for territory (for example, Bosnia), the macrolevel being the struggle of states from different civilizations for power, international institutions, and influence over third parties (for example, the Gulf War or the U.S.-Japan economic struggle for world markets). Huntington sloganizes a new axis of global politics ("the West versus the Rest") and proclaims that the Iron Curtain of ideology in Europe has been replaced by a "Velvet Curtain of culture," which is cartographically displayed in a map of Middle Europe with a line dividing "Western Christianity circa 1500" from "Orthodox Christianity and Islam."[39]

Huntington's concept of a "civilization" is a curious one that is crucial to his writing of global political space. A deterministic totality that is not reducible to either religion, ethnicity, geography, or attitude, his classification gestures to all these factors. The Japanese, Latin American, and African civilizations appear to be specified geographically, whereas the Confucian, Islamic, Hindu, and Slavic-Orthodox civilizations are specified in religious and ethnic terms. Western civilization is somewhat unique in that it has universalistic ambitions and is apparently secular. However, to specify Huntington's thesis in purely representational terms is to miss how the notion of "civilization" functions as a flexible, free-floating sign for Huntington, a sign that refers to other signs and not to any stable referent. To evoke a "civilization" is to call up a foundational identity, a mystical and mythical transcendental presence that is vague yet absolutely fundamental. To designate a conflict a civilizational one is to determine its character in a definitive and totalizing manner. It is to impose a closure upon events, situations, and peoples.

The geographical specificity and place-based particularity of conflicts are reduced to the terms of a civilizational script. In much the same way as Cold War discourse depluralized and homogenized global space, Huntington's civilizational discourse reduces the geographical specificity of conflicts to reified identities and attributes, transforming their ambiguities and indeterminacies into graspable certainties and solid truths. The multiplicity of identities that traverse the world's peoples are diminished to a set of essential differences and distinctions. States are stamped with civilizational labels: Western states, Confucian states, Islamic states, Hindu states, Latin American states, and Orthodox-Slavic states (and sometimes combinations like Islamic-Confucian states). Global space is "civilized."

This civilizing of global space produces, as we might expect, highly problematic interpretations of the world's various conflicts. For example, the conflict in Yugoslavia is located on the fault line between Western Christianity, Orthodox Christianity, and Islam. For Huntington, this location becomes an explanation. But to reduce the war in Bosnia to an ancient fault line civilizational struggle is to read it in the same terms as those who wish to produce it as an essential civilizational war of the Orthodox Slavic Serbs against Islam. The possibilities of "Bosnia" as a multicultural state and "Bosnian" as a multicultural identity are precluded. Huntington, in other words, accepts the Bosnian Serb leadership's interpretation of the war and then cites this as illustrative of his thesis. The multiculturalism of the region's past is ignored and the complexity of its present struggle is reduced to a sight/site/cite of an ahistorical essential antagonism. Huntington does the same for the Gulf War, accepting, as Fouad Ajami noted, Saddam Hussein's interpretation of the conflict as one "between the Arabs and the West."[40] The Gulf War is written as a case of civilization rallying, a kin-country syndrome whereby groups or states belonging to one civilization rally to the support of other members of their civilization when they are involved in wars and conflicts. Such a claim, however, simplifies the multiplicity of different ways in which the Gulf War was understood and interpreted by different groups in different places. In yet another example, Huntington arbitrarily reads U.S.-China relations in civilizational terms, with the end of the Cold War seeing the "reassertion" of "underlying differences" between the United States and China (a vague claim that is dubious, given the Bush and Clinton administrations' reassurances to China). A reported statement by Deng Xiaoping — China's reformist, Marxist leader who supposedly represents alien non-Western "Confucian" values — that a new cold war is under way between both countries is cited as evidence of civilizational clash.

Huntington's civilizing of the deterritorializing space of the post–Cold War unknown is also extended as an explanation of development struggles in certain states. The efforts by governing elites in countries like Turkey, Mexico, Russia, and other states to open up their territorial economies to global markets mark these countries as "torn countries" within Huntington's civilizational tableaux. These are countries where two civilizations — Western and non-Western — are at war for the identity of the society and state. Such reasoning is problematic in a number of ways. First, it assumes that most states are not torn but are isolated entities with stable and fully formed identities. This has never been the case in the history of the modern world system, where the identity of states is a product of ongoing histories of struggle and mutual interaction, not isolationism. Second, it reduces economic transformations to cultural wars. Postmodernity and globalization become *Kulturkampf*, which then itself becomes explanation. The materiality of the cultural and ideological battles Huntington identifies is ignored. The problematic of countries like Turkey, Mexico, and Russia is much more complex than one of elites creating "torn countries" by trying to make their non-Western countries part of the West.

As many commentators have noted, Huntington's thesis is remarkably simplistic and comprehensively flawed. It is significant, nevertheless, as an example of how neoconservative intellectuals of statecraft are endeavoring to chart global space after the Cold War. What is most interesting about this act of geo-power is how it uses the assumptions, goals, and methods of Cold War strategic culture to reterritorialize the global scene in a way that perpetuates the society of security and politics as *Kulturkampf*. We find this in Huntington in three significant ways. First, like Luttwak, Huntington's thesis reterritorializes global space by triangulating from the same ahistorical realism found in Cold War strategic discourse. This charts international politics as a perpetual struggle for power between coherent and isolated units, each seeking to advance their interests in a condition of anarchy. As Richard Rubenstein and Jarle Crocker note, "Huntington has replaced the nation-state, the primary playing piece in the old game of realist politics, with a larger counter: the civilization. But in crucial respects, the game itself goes on as always."[41] This notion of an even more intense playing of a now redundant game is important. In producing a "civilized" global scene, Huntington is creatively seeking to save political realism from the condition of postmodernity by generating a hyperreal civilizational order in the face of the spatial vertigo of the new world disorder. In using terms that are without definitive referents (like "the West" or "the non-West"),[42] Huntington simulates civilizations and an ordered global political scene. In Baudrillard's terms, he produces truth effects that hide the truth's nonexistence.

Second, the purpose of Huntington's post–Cold War strategic discourse is the same as that of Cold War strategic discourse: to perpetuate the primacy of the United States in world affairs. This is to be achieved by the United States renewing its Western civilization from within and actively containing, dividing, and playing off other civilizations against each other. What is different from the Cold War is how the "West" is rewritten by Huntington to partially exclude Japan, the United States' traditional Cold War ally. Japan has an ambivalent status for Huntington. On one hand, he notes, the West faces no economic challenge "apart from Japan," yet, on the other hand, he describes world economic issues as settled by "a directorate of the United States, Germany and Japan, all of which maintain extraordinarily close relations with each other to the exclusion of lesser and largely non-Western countries."[43] The reason for Huntington's ambivalence about the territorial extent of "the West" is that it refers as much to an imaginative and idealized cultural order as it does to a geographical place. Rooted in the political mythology of Western Europe and North America, the "West" is not simply a geographical community but a universalistic creed of individualism, liberalism, constitutionalism, human rights, democracy, and free markets.[44] It is simultaneously a real place and an imaginary cultural order. "The West versus the Rest" is not simply a spatial struggle between a distinct "here" (the West) and an identifiable "there" (the Rest), but a cultural and spatial struggle that occurs *everywhere* (just like the Cold War, for it too, for Huntington, was a clash of civilizations). Huntington's neoconservative anxiety is a product of the emergent disjuncture between the real and the imaginary West brought about by globalization and postmodernity. That the American economy is now

dependent upon Japanese industrial and finance capital and that significant proportions of the U.S. population belong to other civilizations (Latin American, African, Confucian, Japanese, and Islamic)
is a cause for alarm, because an idealized "West" is being weakened
and undermined from within. The struggle between "the West and
the Rest," therefore, begins on the home front in the fight for domestic economic renewal (reducing America's dependence on foreign capital) and against "multiculturalism," which Huntington associates
with "the de-Westernization of the United States." The internal *Kulturkampf* Huntington describes is viewed in apocalyptic terms. "If . . .
Americans cease to adhere to their liberal democratic and European-
rooted political ideology, the United States as we have known it
will cease to exist and will follow the other ideologically defined
superpower onto the ash heap of history."[45] Huntington's reasoning
points to a geo-politics of exclusion. Deterritorializing geographical
space is to be hardened against foreign civilizations and reterritorialized along the lines of an imaginary Euro-American cultural and
political order. The real is to be disciplined to fit the imaginary.

Third, Huntington's thesis is a writing of a world of threats to
the United States, a world of potential and actual cold wars that require a renewal of the society of security within the "West." Interestingly, Huntington's earlier preoccupation with the "economic cold
war" against Japan is not as prominent as before. The new danger
is a "Confucian-Islamic connection" that features a militaristic Chinese economy exporting arms to Islamic states that are determined
to seek nuclear, chemical, and biological weapons capabilities. "A
Confucian-Islamic military connection has . . . come into being, designed to promote acquisition by its members of the weapons and
weapons technologies needed to counter the military power of the
West. . . . A new form of arms competition is thus occurring between
Islamic-Confucian states and the West" (47). This is occurring at a
time when Western states are reducing their military power. Huntington's response, among other things, is to call for a moderation in
this reduction of Western military capabilities and for the West to
"maintain military superiority in East and Southeast Asia" (49).

Like Luttwak's book, Huntington's thesis is interesting not for
its explanatory value but as a document of a certain structure of
feeling within the U.S. foreign policy community. This structure of
feeling is a reactionary one in the sense that it reacts to the vertigo

and complexity of postmodernity with a long-standing conservative pessimism and fundamentalism. As James Kurth points out, the term "Western civilization" was only invented at the beginning of the twentieth century and was itself a sign of a pessimistic feeling of decline within Europe (most pointedly expressed by Oswald Spengler).[46] This pessimism remains at the end of the twentieth century within many elite academic and governmental circles in the United States as it faces an increasingly disorderly world over which it has less and less control. Huntington's partly resigned but also partly defiant response to this unhappy condition is to return to the imaginary fundamentals of earlier history and recycle them in the hope of reterritorializing global space in such a way that his neoconservative agenda of cultural and ideological war against those who would challenge Western fundamentalism (its national security state and society of security) becomes the only option. Huntington's thesis is not about the clash of civilizations. It is about making global politics a clash of civilizations.

EPILOGUE: CONSTELLATIONS OF GEO-POWER AT THE END OF THE TWENTIETH CENTURY

In the maelstrom of the time-space compression of the last fin de siècle, geopolitics congealed as a governmentalized form of geography designed to envision and discipline the spinning globe to a fixed imperial perspective. A floating sign without an essential identity, geopolitics imperfectly named a practice where geographical knowledge was combined with political imperatives by intellectuals of statecraft to envision and script global space in an imperial manner. A century later, a series of new congealments of geography and governmentality are emerging amid an even more intense round of time-space compression, a *fin de* millenium vertigo of informationalization and globalization that is remaking global space and creating new conditions of possibilty for its representation by systems of authority. The challenge for critical geopolitics today is to document and deconstruct the institutional, technological, and material forms of these new congealments of geo-power, to problematize how global space is incessantly reimagined and rewritten by centers of power and authority in the late twentieth century. As should be evident from this chapter and the last, critical geopolitics needs to confront the overdetermining role now played by the media, technology, and the political economy of globalization in enframing the global. While congealments of geo-power are best studied in their specificity, a few final points on the general problematic at the end of the twentieth century are in order.

First, the creation and consolidation of global media networks by transnational media corporations like Disney–Capital Cities, Time-Warner, and News Corp. over the past decade have transformed how we see and experience global politics. Within the circuits, feeds, and flows of networks like CNN (whose icon is a perpetually spinning globe), global political space is skimmed twenty-four hours a day and produced as a stream of televisual images featuring a terrorist attack here, a currency crisis there, and a natural disaster elsewhere. Global space becomes global pace. Being there live is everything. The local is instantly global, the distant immediately close. Place-specific political stuggles become global televisual experiences, experiences structured by an entertainmentized gaze in search of the dramatic and the immediate. Control over the means of televisualizing events (as Ted Turner, Rupert Murdoch, and Silvio Berlusconi well know) is now crucial to the exercise of power and authority in the world. The emissions and power projectionism of the media world a televisual global not only for the masses but also for the new class of rapid reaction global managers whose very jobs depend on their ability to move capital, troops, and technology with the greatest speed possible in response to the latest information. Unlike the situation rooms of old, contemporary situation rooms are full not of maps but of computer screens and television monitors. A considerable part of foreign policy, as we saw in the last chapter, is now preoccupied with the ballistics of televisual projectiles, the trajectory, velocity, and potentially explosive impact of controversial images working their way through media space. Inevitably, geopolitics has itself become a televisual and entertainmentized phenomenon. Mediagenic "wise men" like Henry Kissinger, for example, offer deep thoughts in a sage voice to suitably impressed news anchors while other geopolitical experts war game with studio models and graphic displays. On display here is an important new type of geopolitics, a televisualizing of global politics that deserves careful study by critical geopolitics.

Second, in contemplating the changing nature of geo-power in the late twentieth century, the role of contemporary technoscience in inventing new ways of imaging ourselves and the earth needs to be considered. In radically expanding the range of our vision from the furthest reaches of the galaxy to the smallest fragment of DNA, contemporary corporate technoscience has redefined the limits and dimensions of that which we can now see and map. Military generals who in the past studied maps in order to grasp the length, breadth, and height of the battlefield now study maps of the electromagnetic spectrum. Daily electronic mappings of the surface of the globe are produced by spy satellites, AWACS (Advanced Warning Airborne Command System) planes, stealthy aircraft, and new generations of unmanned aerial vehicles. The organization of the surface of the earth into digitized information stored in geographical information systems (GIS) has technologized geo-power to an unprecedented degree. Places are now digital points, electronic sites made visible by GIS grids and global positioning devices.[47] In the digitized world of cybernetic war machines, everything that can be seen can be destroyed. The challenge is to disappear, to avoid and deflect electromagnetic grids. All of these developments prompted Paul Virilio to proclaim that geopolitics as the management of space has been eclipsed by chronopolitics, the management of time. Space, Virilio provocatively declared, now resides in electronics, not territory.[48] In the military realm, this is indeed the case. Perhaps the most striking example is TERCOM, the terrain contour mapping software brain that guides cruise missiles through hostile territory to their GIS-programmed targets. In this instance, geographical knowledge has come together with governmentality to create a geo-power that is armed, automated, and primed to speed to the kill.

Third, overdetermining the problematic of the writing of global space at the end of the twentieth century is the receding power of the state relative to the global economy in mastering space. Indeed, in many parts of the world, space is no longer mastered by the state at all but by regional warlords and criminal gangs with connections to the global economy. On the African continent, as Robert Kaplan has argued, places like the Sudan, Nigeria, Sierra Leone, Liberia, Zaire, Rwanda, Burundi, and Somalia are nominally states but, in practice, are something else.[49] Most of Angola's diamonds, Cambodia's timber, Peru's cocaine, and Afghanistan's poppy plants are controlled, exploited, and sold to the international market by bandit groups operating beyond the state. Even in industrial and industrializing regions, governmental authority is in retreat in the face of successful contraband economies mediating the global and the local outside the national. In all of the world's states, significant portions of the economy operate outside the law and official statistics. In Russia, for example, organized crime, in partnership with former Communist Party capitalists, controls as much as 40 percent of the turnover in goods and services in the economy.[50] In more affluent regions, global space is also being remastered, as transnational enterprises try to instrumentalize relatively strong states to serve their postnational interests while the wealthy in general try to downsize the state to a self-serving minimalist functionalism. The European Union envisions Europe largely within the terms of a corporatist vision of the Continent as an integrated environment for capital accumulation. In the United States, transnational corporations and their ideologues have successfully rezoned the territory of the United States into a North American free trade region that presently includes Canada and Mexico and will soon incorporate Chile. New transnational corporatist spaces are envisioned for the future. A free trade zone from Alaska to Argentina is promised by the year 2005, while an enormous Pacific Basin free trade zone is projected for the year 2020.

Whatever the nature of new writings of global space by intellectuals of statecraft, all must now take account of the dynamics of informationalization and globalization, of integration and disintegration, in a world where the modern sociospatial triad of the interstate system is in crisis. As already noted, the meaning of state sovereignty is questionable when even the most powerful of states depend upon the goodwill of private financial markets for their economic health and security. Similarly, transcontinental missiles, global satellite television, the Internet, and global warming have rendered territorial integrity a problematic notion. Statist notions of community are also straining, as nominally national but functionally transnational classes use the state to serve their global interests and attempt to secede from common national space by barricading themselves off from the rest in gated enclaves of privilege and affluence. Although somewhat simplistic, the notion of the world being increasingly marked by "wild zones" of poverty and violence, on the one hand, and "tame zones" of wealth and privatized security, on the other hand, is not that farfetched when one considers Washington, D.C., the former capital of the "free world" and current exemplary of a starkly bifurcated world of "haves" and "have nots."[51] The distance between the crack houses and ghettos of the central city, its wild zones of urban poverty and violence, and the master-planned elegance and simulated historic charm of edge cities like Reston, Virginia, the district's outlying postsuburban tame zone where traditional governmental functions like security and zoning are privatized, is only a matter of a few miles, but it is a distance that perhaps foreshadows the spatial structure of twenty-first century transnational corporatist capitalism.[52]

Managing the wild zones of the globe and protecting the security of its tame zones will certainly exercise the minds of the geopoliticians of the future, geopoliticians who will invariably construct their mappings of global space from the standpoint of tame regions and with the agenda of protecting the privileges of the affluent and tame against those who appear to threaten their spatial security and quality of life. Indeed, the recent greening of governmentality marked by the discourse of politicians like Al Gore, the rising influence of think tanks like the aptly named World Watch Institute — a technocratic institution devoted to monitoring the state of the world's environment — and the emergence of a new congealment of geo-power called "environmental security" can be interpreted as a response to the problems that decades of environmental degradation are posing for the rich and powerful, planetary-wide dilemmas involving questions of production, technology, sustainable development, and consumerism that the rich can no longer afford to ignore.[53] Even in their relatively immunized tame zones, the world's richest peoples and ruling classes will be affected. Thus questions of ozone depletion, rainforest cover, biodiversity, global warming, and production using environmentally hazardous materials are the subject of new environmentalist mappings of the global, contemporary acts of geo-power that triangulate global space around the fears and fantasies of the already affluent.

Commenting on these issues, and with Robert Kaplan's vivid description of a coming anarchy in world affairs in mind, President Clinton, in an address to the National Academy of Sciences on sustainable development, revealed a strikingly televisual consciousness of these issues when he remarked that "if you look at what is going on, you could visualize a world in which a few million of us live in such opulence we could all be starring in nightime soaps. And the rest of us look as though we're in one of those Mel Gibson 'Road Warrior' movies."[54] Yet, following the transnational liberalism of his predecessors, President Clinton's plans for embedding the U.S. territorial economy in transnational free trade zones will inevitably hasten this process as significant segments of the U.S. workforce are made redundant by cheaper foreign producers with unimpeded access to the U.S. market.[55] With increasing globalization comes increasing deterritorialization and with increasing deterritorialization comes increasing insecurity. This, in turn, can render the need for the old foundational myths of state, territory, and identity all the greater. In the midst of the unraveling of these old apparent certainties, the will to remythologize them in ever more aesthetic ways can intensify.

Barnet and Cavanagh's declaration at the outset of this chapter—that the fundamental political conflict in the opening decades of the new century will not be between nations or even trading blocs but between the forces of globalization and territorially based forces of local survival seeking to preserve and to redefine community—anticipates this. Yet, at the same time, it does not acknowledge how a backlash by certain "territorially based forces of local survival" can territorialize globalization in such a way as to represent it as conflicts between nations, trading blocs, and civilizations. Seemingly anachronistic identities and dormant territorial disputes between states can take on renewed symbolic meanings amid the dislocations of globalization.

Luttwak's and Huntington's acts of geo-power reveal an anxiety at deterritorialization and a will to reterritorialize global political space that make their work both exemplars yet also complicating refutations of Barnet and Cavanagh's claim. Their work is part of a larger culture of anxiety about the disassembling "West," a culture riven by its contradictory commitments to transnational liberalism and to conservative nationalism. Politicians courting opinion-poll popularity are now simultaneously championing open markets yet closed borders. Leaders are now more aware of transnational threats to the planet yet more willing to fall back on neo-isolationist rhetoric and retreat into a neo-Malthusian stance toward the problems of the world.[56] Fears about the restlessness of the non-Western rest abound in campaigns to harden borders, crack down on immigrants, and "retake" mythic homelands from despoiling foreigners.

Revanchism and xenophobic nationalism are on the rise in the political cultures of relatively affluent states, largely white masculinist cultures of resentment over lost power and diminishing influence, the more extreme versions of which find expression in racist attacks on foreigners or incidents like the Oklahoma City bombing.

Confronting the operation of all these emergent constellations of geographical knowledge and power is a pressing intellectual and political challenge. In taking up this challenge, we would do well to remember that the general problematic of geopolitical discourse — the writing, global space by intellectuals of statecraft — is a complex and messy one that traverses all four substantive sources of social power identified by Michael Mann in his account of the mutual development of classes and nation-states from the eighteenth century.[57] Geopolitical discourses are inevitably entwined with economic sources of power. The imperialist visions of classical geopoliticians, for example, were all shaped by economic interests, materialist motivations, and commercial aspirations, although the saliency of these factors varied considerably, from the relatively insignificant (Ratzel and Haushofer) to the moderate (Mackinder) and strongly significant (Bowman). Geopolitical discourses are also entwined with ideological sources of power. Religious, racial, nationalist, and patriarchal ideologies of identity and difference have all conditioned geopolitical discourse. Yet, while the politics of identity is important to the functioning of geopolitics as an ideology itself, it is not reducible to questions of identity. Geopolitical discourses are furthermore entwined with the growth in political sources of social power, its founding intellectuals owing their very careers to the expanding infrastructural power of the polymorphous nation-state from the late nineteenth century onward. Finally, and perhaps most significantly, geopolitical discourses are entwined with the military as a distinct source of power in modernity. While the military as an institution monopolized a declining relative share of overall state revenues in the nineteenth and twentieth centuries (exempting wartime), it nevertheless expanded in absolute terms, professionalizing and bureaucratizing itself, all the while accumulating greater and greater destructive power.

If geopolitical discourse is organically connected to one social phenomenon above all others in the twentieth century, it is militarism. Militarism, of course, involves not only the military. It is a state-society crystallization that brings together autocratic bureaucracy, technocratic professionalism, class segments, lethal technology, economic interests, and popular nationalism, a crystallization that interweaves elements of many different sources of social power and is reducible to no single factor. As Mann points out, it is a crystallization historically independent of and powerful over all other state crystallizations.[58] This autonomy proved to be fateful on many occasions in the twentieth century beginning in 1914. Yet, profound as the destructive influence of militarism has been on the twentieth century, contemporary geography and political science rarely give it the attention it deserves.

Therefore, in problematizing the re-envisioning of global space at the end of the twentieth century and in developing an agenda for critical geopolitics in the twenty-first century, let us remember that geopolitics is a complex phenomenon embedded in multiple, overlapping networks of power within contemporary states, a phenomenon with long-standing connections to militarism. Geopolitics, as noted in the introduction, is not a singularity but a multiplicity, a twentieth-century constellation of the more general problematic of geo-power, the entwined development of geographical knowledge with the power apparatuses of the modern state. From the cartographic charts of the Elizabethan war machine in Ireland to the electronic grids of the U.S. war machine in the Persian Gulf, geography has long served as a technology of power enframing space within imperial regimes of truth and visibility. Yet the imposition and smooth unfolding of such imperial orders of space has never been without contestation and resistance (as the unfortunate Bartlett found out). Critical geopolitics is one of many cultures of resistance to Geography as imperial truth, state-capitalized knowledge, and military weapon. It is a small part of a much larger rainbow struggle to decolonize our inherited geographical imagination so that other geo-graphings and other worlds might be possible.

Notes

INTRODUCTION: GEO-POWER

1. Sir John Davis (circa 1570–1626) was the English court's solicitor-general for Ireland in 1603; he was knighted for his services (and flattery) in 1607. His *Discovery of the True Causes Why Ireland Was Never Entirely Subdued until the Beginning of His Majesty's Reign* (1612) recommended what would now be called the "ethnic cleansing" of Ireland. This passage is from his correspondence with the Privy Council on difficulties encountered in administrating the plantation of Ulster: "Letter to the Privy Council on Molestation of Geographers, from Coleraine, 23 August 1609," cited in Andrew Hadfield and Willey Maley, "Irish Representations and English Alternatives," in *Representing Ireland: Literature and the Origins of Conflict, 1534–1660*, ed. Brendan Bradshaw, Andrew Hadfield, and Willy Maley (Cambridge: Cambridge University Press, 1993), 13. In it the name "Barkeley" is used. Since the spelling of names was not standardized at this time, this is almost certainly a reference to Richard Bartlett, the English mapmaker who first published a general description of Ulster and ended up being decapitated in county Donegal. See J. H. Andrews, "Geography and Government in Elizabethan Ireland," in *Irish Geographical Studies in Honour of E. Estyn Evans,* ed. Nicholas Stephens and Robin Glasscock (Belfast: Queens University, Department of Geography, 1970), 181.

2. UN High Commissioner for Refugees, *State of the World's Refugees 1993* (London: Penguin, 1993), iii–1.

3. See T. W. Moody, *The Londonderry Plantation, 1609–41: The City of London and the Plantation of Ulster* (Belfast: William Mullen, 1939).

4. Michel Foucault, "Of Other Spaces," trans. Jay Miscowiec, *Diacritics* 16 (Spring 1986): 22–27. See also Henri Lefebvre, *The Production of Space,* trans. Donald Nicholson-Smith (Oxford: Blackwell, 1991), 45.

5. Michael Shapiro, *Reading the Postmodern Polity* (Minneapolis: University of Minnesota Press, 1992), 108–10.

6. R. B. J. Walker, *Inside/Outside: International Relations as Political Theory* (New York: Cambridge University Press, 1993), 125–40.

7. In the heat of an actual campaign, Andrews claims, maps were probably of limited importance. The Irish were not known to have used them, while the English armies probably relied on guides for most of their topographical intelligence. See Andrews, "Geography and Government," 185.

8. Roy Foster, *Modern Ireland, 1600–1972* (London: Penguin, 1988), 35.

9. David Barker, "Off the Map: Charting Uncertainty in Renaissance Ireland," in *Representing Ireland,* ed. Bradshaw et al., 81.

10. According to Cormack, geography in late-sixteenth-century England comprised mathematical geography (geodesy and cartography), descriptive geography (an easily accessible popular geography), and chorography (which included genealogy and chorology as well as local history and topography). While geography was an identifiable discourse, it was not necessarily one separate and distinct from political discourse, history, or even poetry at this time. Julia Reinhard Lupton points to the word "plot" as a key concept that traverses all these genres. It encompasses English strategies for Irish reform, the cartographic projects of surveying and mapping that furthered them, an antiquarian approach to Irish history, and, bringing these all together, the poetics of a figure like Edmund Spenser. See Lesley Cormack, " 'Good Fences Make Good Neighbors': Geography as Self-Definition in Early Modern England," *Isis* 82 (1991): 639–61; Cormack, "Geography and the State in Elizabethan England," in *Geography and Empire,* ed. Anne Godlewska and Neil Smith (Oxford: Blackwell, 1994), 15–30; and Julia Reinhard Lupton, "Mapping Mutability; or, Spenser's Irish Plot," in *Representing Ireland,* ed. Bradshaw et al., 93–115.

11. Among these works are Richard Stanihurst's description of Ireland in Holinshed's *Chronicles* (1577), John Derricke's *Image of Ireland* (1581), Robert Payne's *Brief Description of Ireland* (1589), Edmund Spenser's *View of the Present State of Ireland* (written 1596, published 1633), and Luke Gernon's *Discourse of Ireland* (1620) and *Advertisements for Ireland* (1623). Edmund Spenser's *The Faerie Queen* was pervasively shaped by his experiences in Ireland. He, like Davis, argued that the Irish were a barbaric race that must first be broken by starvation and the sword before they could be remade and reformed. William Petty's classic *The Political Anatomy of Ire-*

land was written around 1672 but not published until 1691. For a discussion of these and other significant medieval writings of Ireland, see the essays in *Representing Ireland,* ed. Bradshaw et al.

12. Foster, *Modern Ireland,* 17.

13. Sir John Davis, November 8, 1610, quoted in Nicholas Canny, *The Elizabethan Conquest of Ireland* (Sussex: Harvester, 1976), 119.

14. This feminization of the Irish landscape is a recurrent theme in the writings of Davis and other contemporaries like Luke Gernon. See Hadfield and Maley, "Irish Representations," 4.

15. For accounts of these encounters, see Tzvetan Todorov, *The Conquest of America* (New York: HarperPerennial, 1982), and Paul Carter, *The Road to Botany Bay* (Chicago: University of Chicago Press, 1987).

16. Michel Foucault, *The Order of Things* (New York: Vintage, 1970); David Cairns and Shaun Richards, *Writing Ireland: Colonialism, Nationalism and Culture* (Manchester: Manchester University Press, 1988), 1–21.

17. Sir John Davis, quoted in Canny, *Elizabethan Conquest,* 135.

18. Michel Foucault, "Governmentality," in *The Foucault Effect: Studies in Governmentality,* ed. G. Burchell, C. Gordon, and P. Miller, trans. C. Gordon (Chicago: University of Chicago Press, 1991), 87–104. The text is a transcribed and edited version of a lecture Foucault gave at the Collège de France in February 1978.

19. Foucault is quoting from Guillaume de La Perriere's anti-Machiavellian book, *Miroir politique* (1567), here.

20. In an interview in 1983, Foucault suggests that territory at the beginning of the seventeenth century came to be envisioned after the model of the city. Cities, and the problems they raised, "served as models for the governmental rationality that was to apply to the whole of the territory. There is an entire series of utopias or projects for governing territory that developed on the premise that a state is like a large city; the capital is like its main square; the roads are like its streets. A state will be well-organized when a system of policing as tight and efficient as that of the cities extends over the entire territory" (259–60). States, in other words, were "urbanized," conceived in the image of the urban. See Michel Foucault, "An Ethics of Pleasure," in *Foucault Live,* trans. Stephen Riggins (New York: Semiotext[e], 1989), 257–77.

21. The case for the significance of Thomas Cromwell in the history of government is made by Geoffrey R. Elton in *The Tudor Revolution in Government: Administrative Changes in the Reign of Henry VIII* (Cambridge: Cambridge University Press, 1953). Michael Mann has challenged this thesis somewhat by pointing out that only after 1660 did the English state's financial size increase substantially in real terms. "In terms of its resource-requiring functions, the Tudor and early Stuart state was late medieval"

(475). Yet, Mann also notes that the permanent-war state dates from the time of Henry VIII. Developments in military logistics, technology, and organization constituted a revolution that favored the most centralized, ordered, and efficiently administered states. To maintain a war machine, a state needed to bolster and augment its revenues. Under Henry VIII, permanent peacetime taxation is introduced for the first time, as is discourse about the general civil benefits of the king's government. See Michael Mann, *The Sources of Social Power,* vol. 1, *A History of Power from the Beginning to A.D. 1760* (Cambridge: Cambridge University Press, 1986), 450–83.

22. Andreas Dorpalen, *The World of General Haushofer* (New York: Farrar and Rinehart, 1942), 48. Turgot planned a proposed book on *la géographie politique* after he had decided to abandon his plans for an ecclesiastical career (he was studying at the Sorbonne Faculty of Theology at the time) in favor of a career of service to the Crown. According to Heffernan, Turgot identified four elements of political geography: "the historical study of the relationships between the physical world, the distribution of population around the globe and the formation of nation states; the study of resources, industries, commerce and wealth within particular countries; the study of transport systems in different countries; and the study of geographical variations in the forms of government and political organization around the world." Heffernan goes on to note that "once acquired, knowledge of these four aspects of political geography should be applied in the service of good government and administration" (334). See Michael Heffernan, "On Geography and Progress: Turgot's *Plan d'un ouvrage sur la géographie politique* (1751) and the Origins of Modern Progressive Thought," *Political Geography* 13 (1994): 328–43.

23. Godlewska notes how statistical inquiry became a universal preoccupation during the revolutionary and Napoleonic periods and how geographers increasingly saw the endeavor as integral to their mission. Napoleon planned to make the Collège de France "a useful appendage of The Great State" and, as part of this drive, planned the establishment of a school of geography at the Collège de France that would be "a central information bureau which would gather information on different parts of the world, trace and record any changes, make these known, and provide guidance to those who sought more information" (39). The school, however, was never instituted. Napoleon's military machine owed some of its success to the work of military geographers, the *ingénieurs-géographes* who produced tens of thousands of maps during Napoleon's reign. The most famous Napoleonic work of "geography" today is probably the twenty-two-volume *Description de l'Egypte* (1800), the significance of which has been discussed by Edward Said and others. See Anne Godlewska, "Napoleon's Geographers (1797–1815): Imperialists and Soldiers of Modernity," in *Geography and Empire,*

ed. Godlewska and Smith, 31–53; Godlewska, "Traditions, Crisis, and New Paradigms in the Rise of the Modern French Discipline of Geography, 1760–1850," *Annals of the Association of American Geographers* 79 (1989): 192–213; Godlewska, "Map, Text, and Image. The Mentality of Enlightened Conquerors: A New Look at the *Description de l'Egypte*," *Transactions, Institute of British Geographers*, n.s., 20 (1995): 5–28; and Edward Said, *Orientalism* (New York: Vintage, 1979).

24. Godlewska notes, for example, that the chief geographer of Napoleon's expedition to Egypt was murdered soon after Napoleon returned to France ("Napoleon's Geographers," 47).

25. Michel Foucault, *Discipline and Punish* (New York: Vintage, 1979), 27.

26. Michel Foucault, *Power/Knowledge* (New York: Pantheon, 1980), 59.

27. Edward Said, "Yeats and Decolonization," in *Remaking History,* ed. Barbara Kruger and Phil Mariani (Seattle: Bay Press, 1988), 10, and *Culture and Imperialism* (New York: Knopf, 1993), 225.

28. The geographical name "Tirconnell" is a crude Anglicization of the Gaelic name, which most probably meant "the land of O'Donnell" (the dominant Gaelic lord in the region). The name was replaced by the current name of the region "Donegal," which reveals the collapse of Gaelic Ireland, for "Donegal" in Gaelic means "the fort of the foreigner."

29. Foster, *Modern Ireland,* 42.

30. John Agnew, "The Territorial Trap: The Geographical Assumptions of International Relations Theory," *Review of International Political Economy* 1 (1994): 53–80.

31. Michael Shapiro makes this very point in a reading of "Guatemala" as a geographic state identity, arguing that to the extent that one accepts and unreflectively reproduces security-oriented geopolitical discursive practices, one engages in implicit acts of recognition of the existing power and authority configurations. See Shapiro, *The Politics of Representation* (Madison: University of Wisconsin Press, 1988), 93.

32. Timothy W. Luke, "Discourses of Disintegration, Texts of Transformation: Re-Reading Realism in the New World Order," *Alternatives* 18 (1993): 229–58.

33. In the Roman Empire, a "province" was a country or territory, outside Italy, under Roman domination. It also meant a sphere of duty and describes an administrative mission as much as it does a geographical region. In speaking of provinces and the provincial, we are already imagining empires and imperial missions. Northern Ireland comprises six out of the nine counties of Ulster. The durability of its description as "the province of Ulster" or simply "the province" in British political culture is indicative of its status as an implicitly colonial place apart from the rest of Britain.

34. Paul Carter's work is a reminder that the map has historically belonged to the discourse of traveling and only gradually became "flat and smooth." It was not a collection of geographical objects imprisoned beneath the grid of latitude and longitude but closer to the picture or to the travel journal. In Carter's terms, Bartlett's work was exploration and journeying, not discovery or survey. See Carter, *The Road to Botany Bay*, 71.

35. Foster, *Modern Ireland*, 5.

36. Mary Hammer, "Putting Ireland on the Map," *Textual Practice* 3 (Summer 1989): 184–201.

37. Said, *Culture and Imperialism*; Seamus Deane, *Celtic Revivals: Essays in Modern Irish Literature, 1880–1980* (London: Faber and Faber, 1985).

38. Said, "Yeats and Decolonization," 13.

39. This struggle continues in contemporary Irish culture. Catherine Nash explores how the Irish artist Kathy Prendergast provokes debate on the links between the gendered body and traditional views of the national landscape in her provocative drawings and sculptures. See Nash, "Remapping the Body/Land: New Cartographies of Identity, Gender, and Landscape in Ireland," in *Writing Women and Space: Colonial and Postcolonial Geographies,* ed. Alison Blunt and Gillian Rose (New York: Guilford, 1994), 227–50.

40. Said, *Culture and Imperialism*, 7.

41. It has been argued, with some justification, that Said's reading of imperialism is too literary. Aijaz Ahmad argues that modern imperialism and the contest over decolonization become mainly a discursive and literary affair for Said, that "imperialism is *mainly* a cultural phenomenon to be opposed by an alternative discourse" (204). Neil Smith adds that there is a "geographical ambivalence in Said: the invocation of geography seems to offer a vital political grounding to Said's textuality until the abstractness of that geography is realized. Virtually absent is any analysis of the historical landscapes depicted in the fiction he discusses" (555). See Aijaz Ahmad, *In Theory: Classes, Nations, Literature* (London: Verso, 1992), and Neil Smith, "Geography, Empire and Social Theory," *Progress in Human Geography* 18 (1994): 550–60.

42. Shapiro, *Reading the Postmodern Polity*, 110.

43. Walker, *Inside/Outside*, 6.

44. John Agnew and Stuart Corbridge, *Mastering Space: Hegemony, Territory and International Political Economy* (London: Routledge, 1995), 49.

45. Simon Dalby, "Critical Geopolitics: Discourse, Difference, and Dissent," *Environment and Planning D: Society and Space* 9 (1991): 274.

46. In using both Foucault and Derrida together, I do not wish to minimize the differences in the projects of both thinkers. I find Foucault's work

useful in elucidating the historical and institutional origins of discursive power formations, while I find Derrida's strategies of reading useful in revealing the aporias within these discursive power formations. Codified contrasts that portray Foucault as a historio-empirical thinker and Derrida as a figure interested overwhelmingly in philosophical issues, however, are inadequate, for they usually underestimate how deconstructionism as a method leads one to pay very close attention to historical, contextual, and genealogical issues. Deconstructionism, in other words, is a historical and not just a philosophical "method."

1. GEOPOLITICS

1. Nicholas Spykman, "Geography and Foreign Policy, II," *American Political Science Review* 32 (April 1938): 236.

2. Thomas Pakenham, *The Scramble for Africa, 1876–1912* (New York: Random House, 1991), xxi.

3. See Horacio Capel, "Institutionalization of Geography and Strategies of Change," in *Geography, Ideology and Social Concern,* ed. David Stoddart (Oxford: Blackwell, 1981), 37–69.

4. David Livingstone, *The Geographical Tradition* (Oxford: Blackwell, 1992), 187–89.

5. Ibid., 189, 190–202.

6. The notion of a natural attitude was first specified by Edmund Husserl. For a discussion of its operation in classical painting, see Norman Bryson, *Vision and Painting: The Logic of the Gaze* (New Haven, Conn.: Yale University Press, 1983). The following section draws heavily on Bryson's discussion.

7. Ibid., 10.

8. Soja claims that not only was spatiality subordinated in the social sciences but that "the instrumentality of space was increasingly lost from view in political and practical discourse." See Edward Soja, *Postmodern Geographies* (London: Verso, 1989), 34.

9. Halford Mackinder, "The Geographical Pivot of History," *Geographical Journal* 23 (1904): 421–44.

10. Thomas Pakenham, *The Boer War* (New York: Random House, 1979), xix.

11. In Mackinder's scheme, only European or European-like powers could effectively occupy space, that is, make space territory. "Civilized powers" territorialized the map of the world; "half-civilized" or savage peoples did not.

12. Immanuel Wallerstein, *The Capitalist World-Economy* (Cambridge: Cambridge University Press, 1979).

13. A worldwide or global view can also be an entire view, a complete, total(izing) view. I use the word "worldwide" rather than "global" here

because it is the term most directly suggested by Mackinder's discourse. The global, we should note, is not a manifest but a produced condition.

14. "To produce is to force what belongs to another order (that of secrecy and seduction) to materialize." See Jean Baudrillard, *Forget Foucault* (New York: Semiotext[e], 1987), 21.

15. Stephen Daniels and Denis Cosgrove, "Spectacle and Text: Landscape Metaphors in Cultural Geography," in *Place/Culture/Representation,* ed. James Duncan and David Ley (London: Routledge, 1993), 58.

16. Lieven de Cauter, "The Panoramic Ecstasy: On World Exhibitions and the Disintegration of Experience," *Theory, Culture and Society* 10 (1993): 1–23.

17. Mackinder, "Geographical Pivot," 439.

18. Benjamin Woolley, *Virtual Worlds* (London: Penguin, 1992), 197.

19. This landpower/seapower conflict has its ancient origins in Europe and gave rise to the different ideas that are at the basis of world civilization. These origins concern the "ancient opposition between Roman and Greek.... The Teuton was civilized and Christianized by the Roman, the Slav by the Greek. It is the Romano-Teuton who in later times embarked upon the ocean; it was the Graeco-Slav who rode over the steppes, conquering the Turanian. Thus modern land-power differs from the sea-power no less in the source of its ideals than in the material conditions of its mobility." Mackinder, "Geographical Pivot," 433.

20. It should be noted that "eastern Europe" in this case is not the "Eastern Europe" of the Cold War imagination but a region stretching from Hamburg and Milan in the west to slightly beyond the Urals in the east.

21. Timothy Mitchell, *Colonizing Egypt* (Berkeley: University of California Press, 1991), and Derek Gregory, *Geographical Imaginations* (Oxford: Blackwell, 1993), 13–69.

22. Gregory, *Geographical Imaginations,* 70.

23. Martin Jay, *Downcast Eyes: The Denigration of Vision in Twentieth-Century French Thought* (Berkeley: University of California Press, 1993), 149–209.

24. Halford Mackinder, *The Teaching of Geography and History: A Study in Method* (London: George Philip, 1918), 40.

25. On Mackinder and modernity, see Gearóid Ó Tuathail, "Putting Mackinder in His Place: Material Transformations and Myth," *Political Geography* 11 (1992): 100–118.

26. Woolf, quoted in Steven Kern, *The Culture of Time and Space, 1880–1918* (Cambridge: Harvard University Press, 1983), 183.

27. Lefebvre, *The Production of Space,* 25. See also David Harvey, *The Condition of Postmodernity* (Oxford: Blackwell, 1989), especially 10–38, 260–83.

28. Livingstone, *The Geographical Tradition*, 197–202; Mark Bassin, "Imperialism and the Nation State in Friedrich Ratzel's Political Geography," *Progress in Human Geography* 11 (1987): 473–95; Paul Weindling, "Ernst Haeckel, Darwinismus and the Secularization of Nature," in *History, Humanity and Evolution*, ed. James R. Moore (Cambridge: Cambridge University Press, 1989), 311–53.

29. Ratzel, quoted in Capel, "Institutionalization," 63.

30. See Woodruff D. Smith, *Politics and the Sciences of Culture in Germany, 1840–1920* (New York: Oxford University Press, 1991), 135–39.

31. On migrationist colonialism, see Woodruff Smith, *The Ideological Origins of Nazi Imperialism* (New York: Oxford University Press, 1986), 21–30. Smith argues that the timing of Ratzel's development of diffusionism strongly suggests that the main direction of influence was from his politics to his scientific theorizing. See his *Politics*, 140–54, especially 152.

32. Friedrich Ratzel, "The Territorial Growth of States," *Scottish Geographical Magazine* 12 (July 1898): 351.

33. Friedrich Ratzel, "The Laws of the Spatial Growth of States," in *The Structure of Political Geography*, ed. Roger Kasperson and Julian Minghi (Chicago: Aldine, 1969), 18.

34. For a discussion of the origins of the concept of lebensraum see Woodruff Smith, "Friedrich Ratzel and the Origins of *Lebensraum*," *German Studies Review* 3 (1980): 51–68. See also his *Ideological Origins* and *Politics*, especially 219–33.

35. Smith, *Politics*, 227.

36. Alfred Thayer Mahan, *From Sail to Steam: Recollections of Naval Life* (New York: Harper, 1907), 276–77; William D. Puleston, *Mahan: The Life and Work of Captain Alfred Thayer Mahan, U.S.N.* (New Haven, Conn.: Yale University Press, 1939), 69; Robert Seager II, *Alfred Thayer Mahan: The Man and His Letters* (Annapolis, Md.: Naval Institute Press, 1977), 145, 431.

37. Seager, *Mahan*, 430.

38. Ibid., 205.

39. Ronald Carpenter, *History as Rhetoric: Style, Narrative, and Persuasion* (Columbia: University of South Carolina Press, 1995).

40. Mahan, quoted in ibid, 108.

41. Seager, *Mahan*, 457.

42. Carpenter, *History as Rhetoric*, 116.

43. Seager, *Mahan*, 291.

44. Ibid, 299.

45. On these general issues in the 1890s, see Robert K. Massie, *Dreadnought* (New York: Ballantine Books, 1991).

46. Michel de Certeau, *The Practice of Everyday Life,* trans. Steven Rendell (Berkeley: University of California Press, 1984), 36.

47. Alfred Mahan, *The Influence of Seapower upon History, 1660–1783* (1890; reprint, New York: Hill and Wang, 1957), 76.

48. This is not to imply that Mahan's particular version of political realism was free from neo-Lamarckian assumptions and reasoning. Sloan quotes Mahan, who in 1897 asserted: "All around us now is strife: the struggle of life, the race of life are phrases so familiar that we do not feel their significance till we stop to think about them. Everywhere nation is arrayed against nation; our own no less than others." See G. R. Sloan, *Geopolitics in United States Strategic Policy, 1890–1987* (New York: St. Martin's Press, 1988), 222; Alfred T. Mahan, *The Interest of America in Sea Power* (London: Sampson Law, 1897), 18. For an example of the theatrical metaphor in Mahan, see *The Story of the War in South Africa, 1899–1900* (New York: Greenwood Press, 1900).

49. Sven Holdar, "The Ideal State and the Power of Geography: The Life-Work of Rudolf Kjellen," *Political Geography* 11 (May 1992): 307–23.

50. Holdar notes that there was even an edition of this book published in braille for German disabled soldiers returning from the war (ibid., 321). Even the blind had access to the geopolitical gaze.

51. Ibid., 316.

52. Johannes Mattern, *Geopolitik: Doctrine of National Self-Sufficiency and Empire* (Baltimore, Md.: Johns Hopkins Press, 1942), 65. Kjellen argued that the new political science must fill "the legal skeleton with social flesh and blood"; it must treat the state as a living organism, acting as a living organism within its own body and in relation to other states as organisms (68–69).

53. Kjellen, quoted in Holdar, "The Ideal State," 312.

54. Ibid., 314.

55. See Geoffrey Stoakes, *Hitler and the Quest for World Domination* (New York: St. Martin's Press, 1986). Although the Haushofer-Hitler relationship was overdramatized, the Haushofer-Hess relationship still generates controversy and mystery. Recently released Soviet files on Hess contain information on his activity in the homosexual underground of Weimar at this time. Hess was, by Haushofer's own admission, his favorite pupil, and he hid Hess in his mountain retreat after the failure of the beer hall putsch. When Hess flew to Scotland in May 1941, he had with him the visiting cards of both Haushofer and his son Albrecht (who apparently knew the duke of Hamilton). It has been alleged that Hess's flight was part of a plan to end the war between Germany and England and to create a joint alliance to fight the Soviet Union. Whatever the truth, Stalin was fed intelligence by Kim Philby that this indeed was the case, thus fueling Stalin's

suspicions of Churchill. A meeting was arranged between Haushofer and Hess in Nuremberg after the war to try to cast light on Hess's mental condition at the time, since he was suicidal and claimed amnesia. Haushofer, however, was unable to jog Hess's memory, and his general behavior caused Haushofer considerable distress. For recent discussions of these issues, see Louis Kilzer, *Churchill's Deception: The Dark Secret That Destroyed Nazi Germany* (New York: Simon and Schuster, 1994), and Ron Rosenbaum, "Kim Philby and the Age of Paranoia," *New York Times Magazine,* July 10, 1994, 29–37, 50, 53–54. For an account of the October 1945 Haushofer-Hess meeting, see Edmund Walsh, *Total Power: A Footnote to History* (New York: Doubleday, 1948), 23–27.

56. Quoted in Donald Norton, "Karl Haushofer and the German Academy, 1925–1945," *Central European History* 1 (March 1968): 80–99, at 83.

57. Translation from Henning Heske, "Karl Haushofer," in *Dictionary of Geopolitics,* ed. John O'Loughlin (Westport, Conn.: Greenwood Press, 1994), 112–13. Different translations are to be found in Andreas Dorpalen, *The World of General Haushofer* (New York: Farrar and Rinehart, 1942), 23; and Robert Strausz-Hupé, *Geopolitics: The Struggle for Space and Power* (New York: Putnam, 1942), 7.

58. Heske, "Karl Haushofer," 139–41.

59. Karl Haushofer, "Defense of German Geopolitics," appendix to Edmund Walsh, *Total Power,* 345.

60. David Murphy, "Space, Race and Geopolitical Necessity: Geopolitical Rhetoric in German Colonial Revanchism, 1919–1933," in *Geography and Empire,* ed. Godlewska and Smith, 173–87, especially 176.

61. Haushofer, "Defense," 346.

62. This meeting took place in Hess's house in Munich. Haushofer reportedly informed Hitler that, after talking to French and English delegates at an international conference on Africa, he had the impression that Germany might be given some territory on the Continent. Germany, however, must make absolutely no moves in Poland. At this Hitler apparently stood up and walked out, never to return. See Sloan, *Geopolitics in United States Strategic Policy,* 38, 54–55.

63. Haushofer, "Defense," 352.

64. Walsh, *Total Power,* 34.

65. Rev. Louis J. Gallagher, S.J., *Edmund A. Walsh, S.J.: A Biography* (New York: Benziger Brothers, 1962), 159.

66. Walsh, *Total Power,* 3–4.

67. Peter F. Coogan, "Geopolitics and the Intellectual Origins of Containment" (Ph.D. diss., University of North Carolina at Chapel Hill, 1991), 336.

68. Gallagher, *Edmund A. Walsh,* 247.

69. David Halberstam, *The Best and the Brightest* (New York: Penguin, 1972), 146–47.

70. For an excellent account of the activities of all of these figures and others, see Coogan, "Geopolitics."

71. Ibid., 165; Fred Kaplan, *The Wizards of Armageddon* (New York: Simon and Schuster, 1983), 19–20; Nicholas J. Spykman, *America's Strategy in World Politics* (New York: Harcourt, Brace, 1942) and *The Geography of the Peace* (New York: Harcourt, Brace, 1944). For a full discussion of Spykman's geopolitical ideas and legacy, see David Wilkinson, "Spykman and Geopolitics," in *On Geopolitics: Classical and Nuclear*, ed. Circo Zoppo and Charles Zorgbibe (Dordrecht: Martinus Nijhoff, 1985), 77–130.

72. Nicholas Spykman, "Geography and Foreign Policy, I," *American Political Science Review* 32 (1938): 29, and *America's Strategy in World Politics*, 41.

73. The ideal of an abstract, universal position, a position beyond race, class, gender, history, and geography, is termed a "view from nowhere" by Bordo, after Thomas Nagel's book of the same name. See Susan Bordo, "Feminism, Post-Modernism and Gender-Skepticism," in *Feminism/Postmodernism*, ed. Linda Nicholson (New York: Routledge, 1990), 137; and Thomas Nagel, *The View from Nowhere* (Oxford: Oxford University Press, 1986).

74. On the "body politic" metaphor, see David Campbell, *Writing Security* (Minneapolis: University of Minnesota Press, 1992), 87–92.

75. See Leslie Hepple, "Metaphor, Geopolitical Discourse and the Military in South America," in *Written Worlds*, ed. Trevor Barnes and Jim Duncan (London: Routledge, 1992), 136–54; and Diana Taylor, "Spectacular Bodies: Gender, Terror, and Argentina's 'Dirty War,'" in *Gendering War Talk*, ed. Miriam Cooke and Angela Woolacott (Princeton, N.J.: Princeton University Press, 1993), 20–40.

76. Halford Mackinder, *Democratic Ideals and Reality* (1919; reprint, New York: Henry Holt, 1942), 2.

77. Spykman, "Geography and Foreign Policy, I," 29.

2. CRITICAL GEOPOLITICS

1. Foucault, *Power/Knowledge*, 149; original emphasis.

2. In the United Kingdom, geographers played vital roles in the Royal Artillery (working on artillery surveying, sound flashing, and flash spotting), the Royal Engineers (charting the Normandy beaches for D-Day, among other things), the Ordinance Survey (satisfying the exorbitant demand for maps at the time), the Special Operations Executive (working on subversive warfare, sabotage, and insurgency), the Hydrological Department (producing hydrographic surveys for logistics strategy and seaborne invasions), and

the Central Interpretation Unit (photographic reconnaissance and photographic interpretation). But perhaps the most remarkable contribution of all was the production from 1941 onward, by supervising geographers at Oxford and Cambridge, of fifty-eight volumes of geographical handbooks on the major theaters of war. The handbooks were designed to provide up-to-date intelligence on the geographical conditions of particular places to aid strategic and tactical operations. Balchin notes that by the end of the war there were few British academic geographers, either in or out of formal war service, who were not involved in some way in the geographical handbook project. See G. V. Balchin, "United Kingdom Geographers in the Second World War," *Geographical Journal* 153 (1987): 159–80.

3. The failure of the discipline of geography to confront its deeply political past was particularly pronounced in Germany, where many of the intellectuals who had worked for the Nazi state and taught German *völkisch* nationalism as *Landschaftskunde* (the study of cultural landscapes) remained in positions of power. For studies of German geography before, during, and after the war, see Michael Fahlbusch, Mechtild Rössler, and Dominik Siegrist, "Conservatism, Ideology and Geography in Germany, 1920–1950," *Political Geography Quarterly* 8 (1989): 353–67; and H. Heske, "German Geographical Research in the Nazi Period: A Content Analysis of the Major Geography Journals, 1925–1945," *Political Geography Quarterly* 5 (1986): 267–81.

4. Yves Lacoste, *La Géographie, ça sert d'abord à faire la guerre* (Paris: Maspero, 1976).

5. My understanding of "postmodern" here is not just the logic of late capitalism, as Fredric Jameson argues, but also as a condition marked by the loss of Eurocentrism and history in their traditional totalizing senses. Postmodernism, Robert Young argues, marks the end of history as a single story; it marks, as Derek Gregory puts it, the "dis-orientation of Occidentalism." See Fredric Jameson, *Postmodernism* (Durham, N.C.: Duke University Press, 1991); Robert Young, *White Mythologies: Writing History and the West* (London: Routledge, 1990), 1–6, 117; and Gregory, *Geographical Imaginations*, 139, 166.

6. Editorial essay, "Political Geography — Research Agendas for the Nineteen Eighties," *Political Geography Quarterly* 1 (1982): 1–18.

7. See K. O. Oye, R. J. Lieber, and D. Rothschild, eds., *Eagle Defiant: U.S. Foreign Policy in the 1980's* (Boston: Little, Brown, 1983); George Black, "Central America: Crisis in the Backyard," *New Left Review* 135 (1982): 5–34.

8. Gearóid Ó Tuathail, "The Language and Nature of the 'New' Geopolitics: The Case of U.S.–El Salvador Relations," *Political Geography* 5 (1986): 73–85; Cynthia Weber, "Shoring Up a Sea of Signs: How the Carib-

bean Initiative Framed the U.S. Invasion of Grenada," *Environment and Planning D: Society and Space* 12 (1994): 547–58.

9. On the teletraditionalism of Reaganism, see Timothy W. Luke, *Screens of Power* (Urbana: University of Illinois Press, 1989), 71–81.

10. Gearóid Ó Tuathail and John Agnew, "Geopolitics and Discourse: Practical Geopolitical Reasoning in American Foreign Policy," *Political Geography* 11 (1992): 190–204. See also Gearóid Ó Tuathail, "Critical Geopolitics: The Social Construction of Space and Place in the Practice of Statecraft" (Ph.D. diss., Syracuse University, 1989).

11. Gearóid Ó Tuathail, "Foreign Policy and the Hyperreal: The Reagan Administration and the Scripting of 'South Africa,' " in *Written Worlds,* ed. Barnes and Duncan, 155–75.

12. Simon Dalby, *Creating the Second Cold War* (London: Guilford, 1990).

13. Dalby, "Critical Geopolitics: Discourse, Difference and Dissent," 274 (see intro., n. 45).

14. See the special issue of *Environment and Planning D: Society and Space* 12, no. 5 (1994) on critical geopolitics.

15. The term "methodogical" is in quotation marks because these methodological principles cannot be considered in any way akin to orthodox scientific procedural rules. As Rodolphe Gasché explains in his chapter on deconstructive methodology, deconstruction is also a deconstruction of the concept of method. I have adopted the notion of "infrastructure" from him. See Rodolphe Gasché, *The Tain of the Mirror* (Cambridge, Mass.: Harvard University Press, 1986), 121–76.

16. See, for example, D. Delaney, "Geographies of Judgement: The Doctrine of Changed Conditions and the Geopolitics of Race," *Annals of the Association of American Geographers* 83 (1992): 48–65; Fredric Jameson, *The Geopolitical Aesthetic: Cinema and Space in the World System* (Bloomington: Indiana University Press, 1992); and H. Savitch, *Post-Industrial Cities: Politics and Planning in New York, Paris, and London* (Princeton, N.J.: Princeton University Press, 1988).

17. E. Thermaenius, "Geopolitics and Political Geography," *Baltic and Scandinavian Countries* 4 (1938): 166.

18. Ibid., 165.

19. R. Schmidt, *Geopolitics: A Bibliography* (Maxwell Air Force Base, Ala.: Air University), 1.

20. David Haglund, "The New Geopolitics of Minerals," *Political Geography Quarterly* 5 (1986): 223.

21. See Saul Cohen, *Geography and Politics in a World Divided,* 2nd ed. (New York: Oxford University Press, 1973), 29–30; and S. Brunn and

K. Mingst, "Geopolitics," in *Progress in Political Geography*, ed. M. Pacione (London: Croom Helm, 1985), 41–47.

22. Geoffrey Parker, *Western Geopolitical Thought in the Twentieth Century* (New York: St. Martin's Press, 1985); Edward Mead Earle, ed., *The Makers of Modern Strategy* (Princeton, N.J.: Princeton University Press, 1943; reprint 1971).

23. Circo Zoppo and Charles Zorgbibe, eds., *On Geopolitics: Classical and Nuclear* (Dordrecht: Martinus Nijhoff, 1985).

24. Sloan, *Geopolitics in United States Strategic Policy*, 25.

25. Gasché, *Tain*, 128.

26. Jacques Derrida, *Of Grammatology*, trans. Gayatri Chakravorty Spivak (Baltimore, Md.: Johns Hopkins University Press, 1976), 19.

27. Gasché, *Tain*, 151–76. Gasché, following Derrida, names the tactic "the logic of paleonymics."

28. Jacques Derrida, *Positions* (Chicago: University Of Chicago Press, 1981), 71; Gasché, *Tain*, 167. The very point of the hyphen is to put geopolitics under erasure, to write geopolitics yet cross it out so as to mark it as a problematic presence. Derrida develops the practice following Heidegger, who expounded on the problem of language having us rather than us having language. The question of Being is anterior to thought and must necessarily be granted and precomprehended in order that thinking can occur in the first place. Derrida questions Heidegger's setting up of Being as a transcendental signified. See Gayatri Chakravorty Spivak, "Translator's Preface," in Derrida, *Of Grammatology*, xiii–xx.

29. See Simon Critchley, *The Ethics of Deconstruction* (Oxford: Blackwell, 1992), 22.

30. De Certeau, *Practice of Everyday Life*, 37 (original emphasis).

31. The contrast between a guerrilla and a surveyor is elaborated by Gregory from the work of Haraway and de Certeau. See Gregory, *Geographical Imaginations*, 160–65.

32. Wlad Godzich, "Foreword: The Tiger on the Paper Mat," in Paul de Man, *The Resistance to Theory* (Minneapolis: University of Minnesota Press, 1986), xiv.

33. Gregory Ulmer, *Applied Grammatology* (Baltimore, Md.: Johns Hopkins University Press, 1985), 36.

34. See James Der Derian, *Antidiplomacy: Spies, Terror, Speed, and War* (Cambridge: Blackwell, 1992), 2; Jay, *Downcast Eyes*, 389.

35. Jay, *Downcast Eyes*, 23.

36. Ibid., 24.

37. Derrida notes how the notion of simultaneity in literary structuralism is associated with the myth of a total reading or description, an argu-

ment that is equally applicable to geopolitics, which frequently presents it-self as an activity that makes the world present in all its parts. It is global in the sense that it is a total(izing) reading.

38. Ibid., 272. See also Ulmer, *Applied Grammatology,* 30–67.

39. Jay, *Downcast Eyes,* 69.

40. See Martin Jay, "Scopic Regimes of Modernity," in *Vision and Visuality,* ed. Hal Foster (Seattle: Bay Press, 1988), 3–28.

41. Yves Lacoste treats Vidal de La Blache's (1845–1918) *Tableau de la Géographie de la France* (1903) as the ur-text of a type of geography that appears natural and descriptive but is actually deeply political. Vidal believed that geography was about the "scientific study of places" and developed an ocularcentric methodology for the discipline in France that involved the presentation of France and its regions as complete legible compositions. See Lacoste, *La Géographie,* 49–59; Gregory, *Geographical Imaginations,* 39; and Livingstone, *Geographical Tradition,* 266–71.

42. Gregory, *Geographical Imaginations,* 13–69. Gregory's account is useful not only in its exploration of ocularcentrism in geography (which includes a contemporary widespread enthusiasm within the discipline for the visual geographical information systems technologies) but also because in so doing he begins to erode the hegemonic post–World War II distinction within the discipline between the figure of the "geographer" and the figure of the "geopolitician." Vidal, for example, long presented as a heroic scientific (that is, nonpolitical) geographer, was actually opposed only to the German school of geopolitics and not to the practice of geopolitics per se. Indeed, his geographical corpus was a form of French geopolitics. On this point, see Michael Heffernan, "The Science of Empire: The French Geographical Movement and the Forms of French Imperialism, 1870–1920," in *Geography and Empire,* ed. Godlewska and Smith, 92–114, especially 103 and 110.

43. For a fuller discussion of this typology, see Gearóid Ó Tuathail, "(Dis)Placing Geopolitics: Writing on the Maps of Global Politics," *Environment and Planning D: Society and Space* 12 (1994): 525–46. For a study of geopolitics in popular culture, see Joanne Sharp, "Publishing American Identity: Popular Geopolitics, Myth, and the *Reader's Digest,*" *Political Geography* 12 (1993): 491–503.

44. See Jacques Derrida, "Signature, Event, Context" in his *Margins of Philosophy* (Chicago: University of Chicago Press, 1982), 307–30; and Critchley, *The Ethics of Deconstruction,* 31–44.

45. See Ronald Carpenter, *History as Rhetoric,* 141–80; and Colin Gray, *The Geopolitics of the Nuclear Era: Heartlands, Rimlands, and the Technological Revolution* (New York: Crane and Russak, 1977).

46. Derrida, *Of Grammatology,* 158.

47. Jacques Derrida, *Limited Inc.* (Evanston, Ill.: Northwestern University, 1988), 153.

48. On the critique of narrow textuality, see Terry Eagleton, *Against the Grain* (London: Verso, 1986), and Edward Said, *The World, the Text and the Critic* (Cambridge, Mass.: Harvard University Press, 1983).

49. Michael Mann, *The Sources of Social Power,* vol. 2, *The Rise of Classes and Nation-States, 1760–1914* (Cambridge: Cambridge University Press, 1993). Late-nineteenth-century geopolitics clearly emerged as part of the expansion of the infrastructural power of the state. The early geopoliticians were all products of the growth of state bureaucracies. Most traversed a series of overlapping social networks of state power. Mahan was born into a small U.S. military establishment in 1840 but died a member of large and powerful institution of American life in 1914. Rooted in the institutions and old regime attitudes of the military, figures like Mahan and Haushofer were able to move into educational institutions and informal circles of government as their careers developed. Others, like Ratzel, Kjellen, and Mackinder, owed their careers as public intellectuals to the growth of state educational institutions; they also moved into political society, sometimes formally serving in national parliaments. Mann describes the geopolitics of the late nineteenth century (geopolitics as science) as helping to systematize the pursuit of "national interests" (746–47). It helped articulate a centralized and territorial conception of interest and community. This conception of "national interests" was inevitably entwined with certain capitalist conceptions of economic interest, but it tended to subordinate economic calculations to territorial and military calculations. In state-societal debates over the specification of the national interest, the "science" of geopolitics came down strongly on the side of militarism and empire.

3. IMPERIAL INCITEMENT

1. Halford Mackinder, "On Thinking Imperially," in *Lectures on Empire,* ed. M. E. Sadler (London: Privately printed, 1907), 38.

2. Some of Mackinder's porters were murdered in skirmishes with inhabitants of the areas they traversed. Mackinder's diary of the expedition was due to be published in 1900 but remained unpublished during his lifetime. Its contemporary editor Michael Barbour speculates on some of the reasons for this. One concerns a note by Campbell Hausberg to the effect that eight Swahili porters had been "shot by orders" during the expedition. He comments that evidence seems to suggest that "something went wrong on the expedition which Mackinder felt could only redound to his discredit if it became widely known." See Michael Barbour, "Editor's Introduction," in H. J. Mackinder, *The First Ascent of Mount Kenya* (London: Hurst, 1991), 23.

3. Halford Mackinder, "A Journey to the Summit of Mount Kenya, British East Africa," *Geographical Journal* 15 (May 1900): 474; Mackinder, *First Ascent*, 220.

4. Brian Blouet, "Sir Halford Mackinder as British High Commissioner to South Russia, 1919–1920," *Geographical Journal* 142 (1976): 228.

5. Joseph Conrad, "Geography and Some Explorers," in *Last Essays*, ed. R. Curle (London: Dent, 1926), 12.

6. Susan Gubar, " 'The Blank Page' and the Issue of Female Creativity," in *The New Feminist Criticism*, ed. Elaine Showalter (New York: Pantheon, 1985), 292–313; Gregory, *Geographical Imaginations*, 129–32.

7. Mary Louise Pratt, *Imperial Eyes: Travel Writing and Transculturation* (London: Routledge, 1992), 29.

8. Ibid., 30.

9. Foucault, *Order of Things*, 128.

10. David Livingstone, "Natural Theology and Neo-Lamarckism: The Changing Context of Nineteenth-Century Geography in the United States and Great Britain," *Annals of the Association of American Geographers* 74 (1984): 9–28, and *The Geographical Tradition*, 63–138.

11. Foucault, *Order of Things*, 25–45.

12. Jay, *Downcast Eyes*, 51.

13. Foucault, *Order of Things*, 130.

14. Ibid., 132.

15. Livingstone, "Natural Theology."

16. Foucault, *Order of Things*, 307–12.

17. Derek Gregory argues that the shift from the classical to the modern *episteme* had two implications for "the protomodern discourses of anthropology and geography." First, it invested geographical space with a linear historicity such that *beyond* Europe became *before* Europe. Second, knowledge as a homogeneous space gave way to knowledge as a three-dimensional space triangulated by mathematics, the empirical sciences, and philosophical reflection. The place of the "protomodern" discourses of anthropology and geography, which, as Gregory suggests, are diffuse and overlapping (remember Ratzel called his geography "anthropogeography"), is a complex and ambiguous one, for, on the one hand, these discourses were structured by the figure of "Man," yet, on the other hand, their inherited natural history methods and the frequently colonial site of their production engaged with an alterity that often undid the monarchy of "Man." See Gregory, *Geographical Imaginations*, 15–33.

18. Pratt, *Imperial Eyes*, 30–34.

19. Mackinder, "Journey to the Summit of Mount Kenya," 454.

20. Carolyn Merchant, *The Death of Nature: Women, Ecology, and the Scientific Revolution* (San Francisco: Harper and Row, 1980), 189.

21. David Stoddart, *On Geography* (Oxford: Blackwell, 1986), 28–40.

22. Mackinder's friend and fellow RGS member George Curzon charged that women's "sex and training render them equally unfitted for exploration, and the genus of professional female globe-trotters with which America has lately familiarized us is one of the horrors of the latter end of the nineteenth century." See Stoddart, *On Geography*; Mona Domosh, "Towards a Feminist Historiography of Geography," *Transactions of the Institute of British Geographers,* n.s., 16 (1991): 97; and the essays in *Writing Women and Space,* ed. Blunt and Rose.

23. E. W. Gilbert, "The Right Honourable Sir Halford Mackinder, 1961–1947," *Geographical Journal* 110 (1947): 97.

24. This description is that of a disenchanted RGS member who did not care for those members of the society who preferred a nonscientific, adventure-travels view of geography to that of more professional researchers. The RGS became increasingly divided between those with an easy aristocratic interest in geography and those with more professional ambitions in the latter half of the nineteenth century (Stoddart, *On Geography,* 59–76). Mackinder's "new geography" of 1885 was sponsored by the latter camp, and he was, therefore, probably keen to shore up his credibility with the socially and politically well connected dilettante circle.

25. Halford Mackinder, "The Development of Geographical Teaching out of Nature Study," *Geographical Teacher* 2 (1904): 192.

26. See Horacio Capel, "Institutionalization of Geography" (see chap. 1, n. 3), and Gerhard Sandner and Mechtild Rossler. "Geography and Empire in Germany, 1871–1945," in *Geography and Empire,* ed. Godlewska and Smith, 115–27.

27. Peter Taylor, "The Value of a Geographical Perspective," in *The Future of Geography,* ed. R. J. Johnston (London: Methuen, 1985), 97.

28. Brian Blouet, *Halford Mackinder: A Biography* (College Station: Texas A & M Press), 3; M. J. Wise, "The Scott Keltie Report 1885 and the Teaching of Geography in Great Britain," *Geographical Journal* 152 (1986): 367–82.

29. For a full discussion of Mackinder's paper, see Paul Coones, *Mackinder's "Scope and Methods of Geography" after a Hundred Years* (Oxford: School of Geography, University of Oxford, 1987).

30. Stoddart, *On Geography,* 77–126.

31. Halford Mackinder, "The Scope and Methods of Geography," *Proceedings of the Royal Geographical Society* 9 (1887): 141–60. All references are to the text *"The Scope and Methods of Geography" and "The Geographical Pivot of History" by Sir Halford Mackinder, reprinted with an introduction by E. W. Gilbert* (London: RGS, 1951).

32. See Blouet, *Mackinder,* 39.

33. Mackinder, "Scope and Methods of Geography," 29.

34. See Ó Tuathail, "Putting Mackinder in His Place."

35. Mackinder, "On the Necessity of Thorough Teaching in General Geography as a Preliminary to the Teaching of Commercial Geography," *Journal of the Manchester Geographical Society* 6 (1890): 4; original emphasis.

36. Mackinder, "On Thinking Imperially," 32.

37. Mackinder provides a more detailed discussion of the difficulties "half-educated" peoples create in democracies in *Democratic Ideals and Reality* (see chap. 1, n. 75). Mackinder contrasts the "organizer" (Bismarck, Napoleon) with the "democrat" in terms of the faculty of sight. The organizer is able to "see things from above." Democracy, however, "implies rule by consent of the average citizen, who does not view things from the hilltops, for he must be at work in the fertile plains" (24).

38. H. J. Fleure, "Sixty Years of Geography and Education: A Retrospective of the Geographical Association," *Geography* 38 (1953): 231–64. It is worth recalling that lantern slides were first perfected by the Jesuits in the seventeenth century as a weapon in a battle of images with Protestantism. The Counter-Reformation generally was intensely ocularcentric and conditioned the deeply visual culture of the baroque. Magic lantern slides were used as visual enhancements in sermons that evoked apocalyptic images of the flames of hell, burning cities, and Armageddon. The images were used not only on congregations but as part of the indoctrination of new Jesuit recruits. The purpose of their use was to seize total control over the participant's imagination. See Manuel De Landa, *War in the Age of Intelligent Machines* (New York: Zone Books, 1991), 187–88; and Jay, *Downcast Eyes*, 42–48. There are many connections between geopolitical thought and religious thought (vision, prophecy, father figures, and so on) that deserve investigation.

39. L. M. Cantor, "The Royal Geographical Society and the Projected London Institute of Geography, 1892–1899," *Geographical Journal* 128 (1962): 30–35.

40. For a detailed discussion of Mackinder and the committee, see James Ryan, "Visualizing Imperial Geography: Halford Mackinder and the Colonial Office Visual Instruction Committee, 1902–1911," *Ecumene* 1 (1994): 157–76.

41. For a discussion of geographical education in France, see Lacoste, *La Géographie*, 41–49. He notes (49) that one text, *Tour de France de deux enfants*, holds a record, after catechism books, for having eight million copies produced since 1877.

42. Foucault, *The History of Sexuality*, 25.

43. Bernard Semmel, *Imperialism and Social Reform* (Cambridge, Mass.: Harvard University Press, 1960).

44. G. R. Searle, *The Quest for National Efficiency* (Oxford: Blackwell, 1971), 64.

45. J. M. Winter, *The Great War and the British People* (London: Macmillan, 1985), 10–15.

46. Searle, *Quest for National Efficiency*, 54.

47. Halford Mackinder, "Man-Power as a Measure of National and Imperial Strength," *National and English Review* 45 (1905): 143.

48. Michael Rosenthal, *The Character Factory: Baden-Powell and the Origins of the Boy Scout Movement* (New York: Pantheon, 1984), 3.

49. Ibid., 192–93.

50. Mackinder, *The Teaching of Geography and History* (see chap. 1, n. 23), 31, and "Development of Geographical Teaching out of Nature Study," 194–95.

51. The pupil is invariably a boy in Mackinder's texts, while the teacher is, beyond the earliest years, normalized as male. There is a curious note in the preface to the second edition of *The Teaching of Geography and History* where Mackinder reflects on the method of address he adopts in the text: "To save space I have often thrown suggestions into an imperative form, and after the first chapter I have referred to the teacher uniformly in the masculine. I make my apologies to those who constitute the majority of the profession" (vi). Mackinder apparently considered the insult of misaddressing a minority of men as women to far outweigh the insult of misaddressing the majority of women as men.

52. Jean-Jacques Rousseau, *Emile*, trans. Barbara Foxley (London: Everyman, 1974), 131.

53. Mackinder, *Teaching of Geography and History*, 3.

54. For a discussion of Starobinski's method, see Jay, *Downcast Eyes*, 86–97.

55. Jean Starobinksi, *Jean-Jacques Rousseau: Transparency and Obstruction*, trans. Arthur Goldhammer (Chicago: University of Chicago Press, 1988), 12.

56. Ibid., 200.

57. Derrida, *Of Grammatology*, 98.

58. Evelyn Fox Keller and Christine Grontkowski, "The Mind's Eye," in *Discovering Reality*, ed. S. Harding and M. Hintikka (Dordrecht: Reidel, 1983), 210.

59. Foucault, *Order of Things*, 319; Jonathan Crary, *Techniques of the Observer* (Cambridge: MIT Press, 1991), 71.

60. See Michel Foucault, *The Birth of the Clinic: An Archaeology of Medical Perception*, trans. A. M. Sheridan Smith (New York: Vintage, 1975).

61. Crary, *Techniques of the Observer*, 95.

62. Halford Mackinder, "The Teaching of Geography from an Imperial Point of View, and the Use Which Could and Should Be Made of Visual Instruction," *Geographical Teacher* 6 (1911): 80; original emphasis. Uppingham was a celebrated public school located in the Midlands. The school had acquired a reputation because of its reformist head, Edward Thring, who died in 1887. Thring was one of the leading proponents of athleticism in the public schools. Game playing promoted "manliness" and made the English a healthy adventurous race. See David Newsome, *Godliness and Good Learning* (London: Cassell, 1961), 220–22; and J. A. Mangan and James Walvin, *Manliness and Morality: Middle-Class Masculinity in Britain and America, 1800–1940* (New York: St. Martin's Press, 1987), 104, 143, 249.

63. Mackinder, *Teaching of Geography and History,* 7.

64. Halford Mackinder, "Comments on Sir Thomas Holdrich's 'The Use of Practical Geography Illustrated by Recent Frontier Operations,'" *Geographical Journal* 13 (1899): 480, and "Comments on W. M. Davis's 'The Systematic Description of Land Forms,'" *Geographical Journal* 34 (1909): 321.

65. Halford Mackinder, "The Human Habitat," *Scottish Geographical Magazine* 47 (1931): 334.

66. Mackinder, "Teaching of Geography from an Imperial Point of View," 81.

67. Ryan points out that many of the photographs taken by Fisher for Mackinder and the Visual Instruction Committee of the Colonial Office were shots taken from a moving train, photographs that proved of little use to Mackinder because they did not always provide a unified vision. Indeed, Ryan suggests that they offer a suitable metaphor for the committee's doomed attempt to impose single photographic meanings upon an imperial world. Ryan, "Visualizing Imperial Geography," 171.

68. Mackinder, *Teaching of Geography and History,* 2. Exactly the same sentiments can be found in *Emile.* Rousseau writes: "The learning of most philosophers is like the learning of children. Vast erudition results less in the multitude of ideas than in a multitude of images" (76 n. 1).

69. Mackinder, *Teaching of Geography and History,* 30.

70. Pratt, *Imperial Eyes,* 7.

71. Derrida, *Of Grammatology,* 101–40.

72. See Mackinder, "Progress of Geography in the Field and in the Study during the Reign of His Majesty King George the Fifth," *Geographical Journal* 86 (1935): 1–12, and "Geography, an Art and a Philosophy," *Geography* 28 (1942): 595–605.

73. Mackinder, "Development of Geographical Thinking out of Nature Study," 193, and *Teaching of Geography and History,* 11–13.

74. A tellurion is a simple kind of orrery used in illustrations of the rotation of the earth and its effects on daylight and the seasons.

75. Mackinder, "Modern Geography, German and English," *Geographical Journal* 6 (1895): 376.

76. Mackinder, "Teaching of Geography from an Imperial Point of View," 81; original emphasis.

77. Mackinder, "Geography, an Art and a Philosophy," 124.

78. David Wood, *The Power of Maps* (New York: Guilford, 1992), 95–142.

79. Brian Harley, "Deconstructing the Map," in *Written Worlds,* ed. Barnes and Duncan, 231–47.

80. Gilles Deleuze, *Foucault,* trans. Sean Hand (Minneapolis: University of Minnesota Press, 1988), 44.

81. Luce Irigaray, *Speculum of the Other Woman* (Ithaca, N.Y.: Cornell University Press, 1985), 144–45.

82. Blouet, *Mackinder,* 153.

83. Jay, *Downcast Eyes,* 211–62.

4. "IT'S SMART TO BE GEOPOLITICAL"

1. Joseph Thorndike Jr., "Geopolitics," *Life* 13 (December 21, 1942): 106–15.

2. For use of the title "Major General Professor Doktor," see Robert Strausz-Hupé, "Geopolitics," *Fortune* 24, no. 5 (November 1941): 111.

3. Deborah Lipstadt, *Beyond Belief: The American Press and the Coming of the Holocaust,* 1933–1945 (New York: Free Press, 1986), 126.

4. Joseph Thorndike, "Geopolitics," 106. The description of geopolitics as a "five dollar term" by *Life* is quoted in Hans Weigert's "Military Implications of German Geopolitics," *Vital Speeches* 8 (August 15, 1942): 645–47.

5. Robert Strausz-Hupé, "It's Smart to Be Geopolitical," *Saturday Review of Literature,* February 6, 1943, 4. The title of Strausz-Hupé's review was akin to an advertising slogan. The February 15, 1943, issue of the *New Republic* carried the following advertisement (213): "THE SATURDAY REVIEW OF LITERATURE says: "It's Smart to Be Geopolitical" — but are you sure you know what GEOPOLITICS means — especially to America?" The books Strausz-Hupé reviewed were all published by Farrar and Rinehart of New York.

6. Describing the question of geopolitics in sales language was not difficult for Strausz-Hupé, since he earned his living in the late thirties giving lecture tours throughout the United States on such topics as "the rape of Austria." In his autobiography, *In My Time* (New York: Norton, 1965), he remarks that his lecture agent advised him on how to pitch his "message" to middle-class audiences (176).

7. Robert Strausz-Hupé, *Geopolitics: The Struggle for Space and Power* (New York: Putnam, 1942). Strausz-Hupé describes the book as a "minor best seller" in his autobiography (*In My Time*, 188). For a critical and preliminary assessment of the career of Robert Strausz-Hupé, see Andrew Crampton and Gearóid Ó Tuathail, "Institutions, Intellectuals, and Ideology: Robert Strausz-Hupé and American Geopolitics," *Political Geography* 15 (1996).

8. Michel Foucault, *The Archaeology of Knowledge* (New York: Pantheon, 1972), 40–49.

9. From Edmund Walsh's account it is clear that the Allied War Crimes Commission spent some time considering Haushofer's relationship to the Nazis and whether to prosecute him at Nuremberg. Haushofer's "influence" on Nazi policy was diagrammed by them, but he was never brought to trial. See Edmund Walsh, "The Mystery of Haushofer," *Life*, September 16, 1946, 106–20, and *Total Power*. For a discussion of the role of the concept of geopolitics in the U.S. governmental bureaucracy during the war, see Coogan, "Geopolitics and the Intellectual Origins of Containment" (see chap. 1, n. 66).

10. Schmidt, *Geopolitics: A Bibliography*; Peter Enggass, *Geopolitics: A Bibliography of Applied Political Geography* (Monticello, Ill.: Vance Bibliographies, 1984).

11. For a summary, see Neil Smith, "Geography Redux? The History and Theory of Geography," *Progress in Human Geography* 14 (1990): 547–59. See also the various relevant entries in John O'Loughlin, ed., *Dictionary of Geopolitics* (Westport, Conn.: Greenwood Press, 1994).

12. Two of the most intellectually sophisticated books published on *Geopolitik* at this time were Johannes Mattern's *Geopolitik: Doctrine of National Self-Sufficiency* and Andrew Gyorgy's *Geopolitics: The New German Science* (Berkeley: University of California Press, 1944). As products of university presses, both works can be considered specialist narrations of geopolitics. They are not considered here because they had a limited impact on shaping the popular conception of geopolitics in the United States, a conceptual space that was already well defined by 1944.

13. *Life*, November 20, 1939, 62–66.

14. *Reader's Digest*, June 1941, 23–27.

15. The *Life* article, which draws upon the disaffected Nazi Hermann Rauschning's *The Revolution of Nihilism* (New York: Alliance, 1939), refers to Haushofer's German Academy, not to a geopolitical institute. The earliest reference I have found to an Institute for Geopolitik is in a comment by the British news magazine the *New Statesman and Nation*, August 26, 1939, 301–2, which reads the Nazi-Soviet pact as the work of Haushofer's adaptation of Mackinder's ideas.

16. The fact that Operation Barbarossa was not seen as a repudiation of Haushofer's ideas is indicative of how flexible and vague the whole Haushofer-as-threat narrative was. Although committed to the thesis that German geopolitics exercised a significant influence on Nazi foreign policy, Hans Weigert comes closest to appreciating the radical disjuncture that Barbarossa marked. "We may assume...that the war on Russia must have been a tremendous shock not only to Haushofer the geographer, but also to Haushofer the soldier." Weigert, German Geopolitics, America in a World at War, no. 19 (New York: Oxford University Press, 1941), 18.

17. "Geopolitics in College," Time, January 19, 1942.

18. Leonard Engel, "Geopolitics and Today's War," Infantry Journal, May 1942, 45.

19. Other examples of popular readings of geopolitics include the article "Persons and Personages: Author of Lebensraum," reproduced in Living Age 359, no. 4492 (January 1941): 434–38, from an unattributed article on the Zurich Independent Weekly Weltwoche; the Newsweek article "Nazi War Scheme: Hitler's Dreams Based on Plan of Englishman," February 17, 1941, 24–28; Joseph C. Harsch, "The 'Unbelievable' Nazi Blueprint," New York Times Magazine, May 25, 1941; and Joseph C. Roucek, "German Geopolitics," Journal of Central European Affairs, April 1942, 180–89.

20. See Thomas Doherty, Projections of War: Hollywood, American Culture and World War II (New York: Columbia University Press, 1993), 16–35.

21. A copy of The Nazi Strike is available from the Film and Stills Division of the U.S. National Archives and Records Administration in Maryland.

22. On Hollywood during World War II, see Clayton Koppes and Gregory Black, Hollywood Goes to War (Berkeley: University of California Press, 1987).

23. Doherty, Projections of War, 78–84.

24. Hans Weigert, "Maps Are Weapons," Survey Graphic, October 1941, 528–30; Weigert, "German Geopolitics: Workshop for Army Rule," Harper's Magazine 183 (November 1941): 586–97; Weigert, German Geopolitics; Weigert, "Military Implications"; Weigert, Generals and Geographers: The Twilight of Geopolitics (New York: Oxford University Press, 1942).

25. Weigert, German Geopolitics, 5.

26. Weigert, Generals and Geographers, 11.

27. Ibid., 8.

28. Weigert, German Geopolitics, 26.

29. Ibid., 29.

30. Weigert, "Military Implications," 646.

31. Strausz-Hupé, "Geopolitics," 111–19; Strausz-Hupé, Axis America: Hitler Plans Our Future (New York: Putnam, 1941); Strausz-Hupé, Geopolitics.

32. Strausz-Hupé, "Geopolitics," 111.

33. Strausz-Hupé, "It's Smart," 5.

34. Strausz-Hupé, "Geopolitics," 114, and *Geopolitics,* 77.

35. Strausz-Hupé, "Geopolitics," 111.

36. This is not to suggest that Strausz-Hupé does not make much of the figure of Haushofer, as we will see. *Geopolitik,* Strausz-Hupé declared, is the product of many minds, but Haushofer is its prophet. Strausz-Hupé, *Geopolitics,* 52.

37. Dorpalen, *World of General Haushofer,* 13.

38. Derwent Whittlesey, *German Strategy of World Conquest* (New York: Farrar and Rinehart, 1942), 3.

39. Many of the books of the early forties are so stamped. The symbol accompanying the text is an eagle poised on three book volumes.

40. For example, see Weigert's *German Geopolitics,* 15, and *Generals and Geographers,* 11.

41. Weigert, *German Geopolitics,* 26.

42. In his article on maps as weapons, Weigert argues that "Hitler's propagandists can teach us a lesson when we attempt to understand the importance of the map as a weapon of propaganda as well as of education." Weigert, "Maps Are Weapons," 528.

43. Frederick Schuman, "Let Us Learn Our Geopolitics," *Current History,* May 1942, 161–65.

44. Ibid., 164. Thorndike's account of geopolitics as a "scientific system which a Briton invented, the Germans used and Americans need to study" notes: "To condemn geopolitics because of the sins of the German geopoliticians is as irrational as to condemn carving knives because they have, on occasion, been used to stab people." Geopolitical knowledge and analysis, he concludes, may be used for an evil purpose, as the Germans used it, but it can also be used to establish and defend a moral order in the world. Thorndike, "Geopolitics," 115.

45. Pierre Van Passen, *The Time Is Now!* (New York: Dial Press, 1941), 20. Van Passen's argument is unusual in that he reads the Geopolitical Institute as sponsoring a plan for war of the oceans, which would give the German state control over the lifelines of the world. This reading of German geopolitics as sea-power geopolitics goes against the grain of all other accounts, which emphasize the land-power dimensions of Haushofer's ideas.

46. Werner Cahnman, "Methods of Geopolitics," *Social Forces* 21 (1942): 152.

47. Strausz-Hupé, "Geopolitics," 119, and *Geopolitics,* 111–12, 86. Although German geopolitics was a threat, Strausz-Hupé discounted the contribution of the institute and *Geopolitik* to German military intelligence.

48. In an article in the *New Republic,* August 10, 1942, 163–64, Frederick Schuman makes explicit the need for a global strategy on the part of the Allies. "The science of global strategy," he notes, "consists in the employment of the tools of war in successive campaigns planned and executed in accordance with a sound geopolitical calculus." Geopolitics, in other words, is instrumental reasoning. It is scientific, technical, and mathematical.

49. Eric Archdeacon to Sloan Colt, quoted in Coogan, "Geopolitics and the Intellectual Origins of Containment," 197.

50. Eric Archdeacon, "Proposed Organization of the American Institute of Political Geography, with Headquarters in Washington, D.C.," quoted in ibid, 198.

51. "The Trend: The Case for Geopolitics," *Business Week,* July 11, 1942, 88.

52. "The Trend: The Case for Geopolitics," *Business Week,* August 1, 1942, 68. This editorial was prompted by an article in *Foreign Commerce Weekly,* a publication of the U.S. Department of Commerce, that reviewed the literature on Germany's Munich Institute, describing it as a research organization whose business was to collect world information, and argued that the Bureau of Foreign and Domestic Commerce of the Department of Commerce had been serving just such a role for America since 1820: "The Bureau is recognized as the best-informed organization of its kind in the world.... It has long filled and continues to fill the role of America's geopolitical institute." *Business Week,* however, was unconvinced, declaring that the United States still had no institute of geopolitics. See Carter R. Bryan, "America's Geo-Political Institute," *Foreign Commerce Weekly* 7, no. 6 (May 16, 1942): 3, 32–33.

53. The Geopolitical Section's mandate, quoted in Coogan, "Geopolitics and the Intellectual Origins of Containment," 202.

54. Ibid, 205.

55. Cahnman, "Methods of Geopolitics," 147–54.

56. Whittlesey, *German Strategy,* 122.

57. Weigert, *Generals and Geographers,* 15–16.

58. Hans Speier, "Magic Geography," *Social Forces* 8 (1941): 310–30. For a revisionist discussion of the *Geopolitik* cartography, see Guntram Herb, "Persuasive Cartography in *Geopolitik* and National Socialism," *Political Geography Quarterly* 8 (1989): 289–303.

59. Ibid., 18.

60. Cahnman quotes Haushofer, who states that the "creative power of an artist" should be combined with the "thorough training of a scholar." A conscientious geopolitician is one who steps "along a narrow mountain path, where he must avoid falling into the vague fantasies of a mystic on

the one hand and the 'psychic impotence and dryness of a taskmaster' on the other." Cahnman, "Methods of Geopolitics," 153.

61. Ibid., 19.

62. Weigert, "Military Implications," 646.

63. Strausz-Hupé, *Geopolitics,* 48.

64. Ewald Schnitzer, "Geopolitics as an Element in Social Education," *Harvard Educational Review* 8 (1938): 507–16.

65. Whittlesey, *German Strategy,* 262.

66. Weigert, *German Geopolitics,* 25.

67. Max Horkheimer and Theodor Adorno, *Dialectic of Enlightenment* (New York: Continuum, 1988), 3.

68. Martin Jay, *The Dialectical Imagination* (Boston: Little, Brown, 1973), 260.

69. Spykman's book *America's Strategy in World Politics* is a good example of how geopolitics was grafted onto an already existent Realpolitik tradition. There are surprisingly few references to "geo-politics" in Spykman's text. These appear in the introduction and in the conclusion, which suggests that they were added late. Furthermore, what the index indicates as references to geopolitics are, in all cases, references to the adjective "geo-political" (as in isolationism and interventionism as two geopolitical schools of thought in the United States). In a useful review of Spykman's book in the *New Republic* 106 (April 20, 1942): 546–47, Malcolm Cowley noted how Spykman's book was being marketed to the U.S. public as "the first comprehensive geopolitical analysis of the position of the United States in the world" (a quote he presumably takes from the original dust jacket). Cowley goes on to remark: "Since geopolitics is now a magical and incantatory word; since everyone is interested in geopolitics as being—perhaps—the secret of Hitler's victories, one reads the book for whatever light it may cast on this new science." This light, he notes, is by no means dazzling. What distinguishes the geopolitical method, he concludes, is not what it includes (it presents very little to "the public store of knowledge about international affairs"), but "rather what it tries to omit—human institutions, human hopes and fears and judgements of right and wrong. Geopolitics might be defined as an attempt to transform the art of international relations into a physical science by excluding everything qualitative; by including nothing that cannot be measured in terms of mass or energy."

70. The story of Renner's maps ("Maps for a New World," *Colliers,* June 6, 1942, 14–16, 28) and the response they provoked from the columnist Walter Lippmann is told in Karen De Bres, "George Renner and the Great Map Scandal of 1942," *Political Geography Quarterly* 5 (1986): 385–94. The story is interesting in that it reveals the depth of the antipathy and sus-

picion that greeted "geopolitics" in the United States. Defining what was geopolitical and what was not, of course, was half the battle. Powerful geographers in the United States like Isaiah Bowman, who incidentally employed Strausz-Hupé on the strength of his book *Geopolitics* in 1942, practiced geopolitics while repudiating the name.

71. Norman Cousins's essay "Generalities and Geopolitics" in the *Saturday Review of Literature* (June 13, 1942, 3–4, 15–16) is an example of the use of geopolitics to describe the one-world vision of Vice-President Henry Wallace. The usage is significant in that, first, geopolitics is discussed separately from any discussion of German geopolitics and, second, it is used to describe an explicitly idealist foreign policy philosophy. Geopolitics, in other words, did not automatically mean Realpolitik. Wallace would later oppose "geopolitical" readings of the Soviet Union as threat after World War II.

72. Horkheimer and Adorno, *Dialectic of Enlightenment,* 20.

73. In 1945 Robert Strausz-Hupé published a survey of world politics, the last chapter of which begins with the sentence: "This book has been written on the assumption that a third World War within this generation, though neither foreordained nor probable, is none the less possible" (256). "Sir Halford Mackinder's concept of the 'heartland,'" Strausz-Hupé concluded, "is today no less than when it was first presented in 1904, the fundamental axiom of world politics" (262). Strausz-Hupé would make a very successful career out of such homilies, a career in which he served as a foreign policy adviser to Barry Goldwater and as U.S. ambassador to Sri Lanka, Sweden, Belgium, NATO, and Turkey for the Nixon, Ford, and Reagan administrations. See Strausz-Hupé *The Balance of Tomorrow* (New York: Putnam, 1945), and Crampton and Ó Tuathail, "Institutions, Intellectuals, and Ideology."

5. CRITICAL APPROACHES TO "GEOPOLITICS"

1. Foucault, *Power/Knowledge,* 66.

2. Karl Wittfogel, "Geopolitics, Geographical Materialism and Marxism," trans. G. L. Ulmen, *Antipode* 17 (1985): 21–72.

3. Jay, *The Dialectical Imagination,* 15.

4. Alan Bullock, *Hitler and Stalin: Parallel Lives* (New York: Knopf, 1992), 167; Ruth Fischer, *Stalin and German Communism* (New Brunswick, N.J.: Transaction Books, 1982).

5. Wittfogel, "Geopolitics," 22.

6. Gary Ulmen, *The Science of Society: Towards an Understanding of the Life and Work of Karl August Wittfogel* (The Hague: Mouton, 1978), 48.

7. Ulmen, *Science,* 92.

8. For a different interpretation and critique of Wittfogel's reading of Marx, see Neil Smith, "Rehabilitating a Renegade? The Geography and Politics of Karl August Wittfogel," *Dialectical Anthropology* 12 (1987): 127–36.

9. Bullock, *Hitler and Stalin,* 13–16, 185, 193, 421.

10. H. Heske, "Karl Haushofer: His Role in German Geopolitics and in Nazi Politics," *Political Geography Quarterly* 6 (1987): 135–44.

11. Wittfogel noted his own misgivings about this policy and that within the German working-class movements at the time. He describes himself as in a "horrible situation because I did not want to break with the Party." See Matthias Greffrath, Fritz Raddatz, and Michael Korzec, "Conversations with Wittfogel," *Telos* 43 (1980): 154.

12. Ulmen, *Science,* 540.

13. Bassin, "Race Contra Space," 134; Bassin, "Geographical Determinism in Fin-de-Siecle Marxism: Georgii Plekhanov and the Environmental Basis of Russian History," *Annals of the Association of American Geographers* 82 (1992): 17–18; and Bassin, "Reductionism Redux? or the Convolutions of Contextualism," *Annals of the Association of American Geographers* 83 (1993): 165.

14. Ulmen, *Science,* 540.

15. Neil Smith, "Bowman's New World and the Council on Foreign Relations," *Geographical Review* 76 (1986): 448.

16. Bowman quoted in ibid., 451.

17. Isaiah Bowman, *The New World: Problems in Political Geography,* 4th ed. (New York: World Book, 1928), iii.

18. Smith, "Bowman's New World," 441.

19. Neil Smith, "Isaiah Bowman: Political Geography and Geopolitics," *Political Geography Quarterly* 3 (1994): 73.

20. Isaiah Bowman, "Geography vs. Geopolitics," *Geographical Review* 32 (1942): 646; original emphasis.

21. Isaiah Bowman, "Political Geography of Power," *Geographical Review* 32 (1942): 350.

22. P. Vigor, "The Soviet View of Geopolitics, " in *On Geopolitics: Classical and Nuclear,* 131–40; O. V. Vitkovskiy, "Political Geography and Geopolitics: A Recurrence of American Geopolitics," *Soviet Geography: Review and Transaction* 22 (1980): 586–593.

23. Smith, "Isaiah Bowman," 73.

24. Neil Smith, "Shaking Loose the Colonies: Isaiah Bowman and the 'De-Colonization' of the British Empire," in *Geography and Empire,* ed. Godlewska and Smith, 270–99.

25. See M. Rössler, "Applied Geography and Area Research in Nazi Society: Central Place Theory and Planning, 1933 to 1945," *Environment and Planning D: Society and Space* 7 (1989): 419–31. Zygmunt Bauman

notes generally how the mystique of white coats and scientific objectivity helped exonerate German experts, now deployed in the service of the victors, from responsibility for their wartime deeds and attitudes. See Bauman, *Modernity and Ambivalence* (Cambridge: Polity, 1993), 42.

26. Between 1940 and 1944 Christaller worked in Heinrich Himmler's Planning and Soil Office under the agronomist Konrad Meyer. This office produced the *Generalplan Ost* (General plan of the east) for Himmler; it proposed a settlement pattern, the construction of new central places (special villages), and a time scale for this work. The plan was never implemented, and Meyer was prosecuted at Nuremberg. Christaller wrote a "whitewash-paper" (apologist testimony) stating that Meyer had employed several scientists who were antifascist in inclination; it was also claimed that the work was purely "scientific." Meyer was freed in 1948. See Rössler, "Applied Geography," 427.

27. Jeffrey Herf, *Reactionary Modernism: Technology, Culture, and Politics in Weimar and the Third Reich* (Cambridge: Cambridge University Press, 1984).

28. See John Dower, *War without Mercy* (New York: Pantheon, 1986), and Alan Henrikson, "The Map as an 'Idea': The Role of Cartographic Imagery during the Second World War," *American Cartographer* 2 (1975): 19–53.

29. Hartshorne letter quoted in Andrew Kirby, "What Did You Do in the War, Daddy?" (paper presented at a conference, "Geography and Empire," Queen's University, Kingston, Canada, April 1991). A shorter version of Kirby's paper with the same title is published in *Geography and Empire*, ed. Godlewska and Smith, 300–315. It omits the conference paper's discussion of Hartshorne in Austria.

30. Richard Hartshorne, "The Functional Approach to Political Geography," *Annals of the Association of American Geographers* 40 (1950): 104.

31. Alan Wolfe, *The American Impasse: The Rise and Fall of the Politics of Growth* (Boston: South End Press, 1984).

32. For the same impulse in *The Nature of Geography*, see Neil Smith, "Geography as Museum: Private History and Conservative Idealism in *The Nature of Geography*," in *Reflections on Richard Hartshorne's The Nature of Geography*, ed. Nicholas Entrikin and Stanley Brunn, Occasional Publications of the Association of American Geographers (Washington, D.C., 1989), 89–120. See also John Paul Jones III, "Making Geography Objectively: Ocularity, Representation, and *The Nature of Geography*," in *Objectivity and Its Other*, ed. Wolfgang Natter, Theodore Schatzki, and John Paul Jones III (New York: Guilford, 1995), 67–92.

33. See Hans Morgenthau, *Politics among Nations* (New York: Knopf, 1949), 116–20.

34. H. Sprout and M. Sprout, "Geography and International Politics in an Era of Revolutionary Change," *Journal of Conflict Resolution* 4 (1960): 145–61, and "Environmental Factors in the Study of International Politics," in *International Politics and Foreign Policy,* ed. J. N. Rossenau (New York: Free Press, 1969), 41–56.

35. S. B. Jones, "Global Strategic Views," *Geographical Review* 44 (1956): 492–495; Cohen, *Geography and Politics in a World Divided.*

36. A. Abdel-Malek, "Geopolitics and National Movements: An Essay on the Dialectics of Imperialism," in *Radical Geography,* ed. R. Peet (London: Methuen, 1977), 263–92.

37. Richard Peet, "The Development of Radical Geography in the United States," in *Radical Geography,* ed. Peet, 7.

38. See Jean-Michel Brabant, Beatrice Giblin, and Maurice Ronai, "Postface," in Lacoste, *La Géographie,* 181–87; and P. George, R. Guglielmo, B. Kayser, and Y. Lacoste, *La Géographie active* (Paris: Presses Universitaires de France, 1964). This later work contrasts an "applied geography" that is dependent on political power to a critical, activist geography that is independent of political power. The radicalization of French geography in the sixties was partly a product of the country's status as a colonial power. Instead of writing the geography of colonial areas in the image of France, French geographers began to use the revolutionary ideologies of anticolonial movements to rewrite the geography and self-image of France.

39. Yves Lacoste, "An Illustration of Geographical Warfare: Bombing of the Dikes on the Red River, North Vietnam," in *Radical Geography,* ed. Peet, 244–61.

40. Lacoste, *La Géographie,* 19.

41. Ibid., 7; I have used the translation of Paul Buleon, "The State of Political Geography in France in the 1970s and 1980s," *Progress in Human Geography* 16 (1992): 27. Subsequent text references are to this translation.

42. Lacoste, "An Illustration of Geographical Warfare," 245.

43. After May 1968, as part of the educational reforms introduced by the government, the Centre Experimental de Vincennes was established and an orientation commission that included George Canguilhem, Emmanuel Le Roy Ladurie, Roland Barthes, and Jacques Derrida was charged with assembling a faculty for the experimental university. Among the faculty they chose was Michel Foucault to head the philosophy department. During its first years of operation, Vincennes was the site of many educational protests and faculty and student confrontations with the police. Foucault spent two years at Vincennes, from December 1968 to December 1970, until elected to the Collège de France. See Didier Eribon, *Michel Foucault,* trans. Betsy Wing (Cambridge, Mass.: Harvard University Press, 1991), 201–11.

44. Brabant, Giblin, and Ronai, "Postface," in Lacoste, *La Géographie*, 185–87.

45. *Hérodote*'s early influences were clearly more Marxist than Foucauldian. Although Lacoste talks about knowledge in general and geography in specific as a form of power in *La Géographie*, his arguments are not developed in depth and appear to draw inspiration from Bachelard (with his notion of ruptures and epistemological breaks) and Althusserian Marxism (with its emphasis on the state's ideological apparatus) above all. Foucault is not mentioned in the text.

46. Michel Foucault, "Questions on Geography," in *Power/Knowledge*, 77. The original French is "Questions à Michel Foucault sur la géographie," *Hérodote* 1 (1976).

47. Yves Lacoste, "Geography and Foreign Policy," *SAIS Review* 4 (1984): 214.

48. Yves Lacoste, "Geographers, Action and Politics," in *International Geopolitical Analysis: A Selection from* Hérodote, ed. and trans. P. Girot and E. Kofman (London: Croom Helm, 1987), 3.

49. Yves Lacoste, "Editorial: D'Autre géopolitiques," *Hérodote* 25 (1982): 5–6.

50. Lacoste, "Geography and Foreign Policy," 225.

51. Lacoste, "Geographers, Action and Politics," 6.

52. Lacoste, "Editorial: D'Autres geopolitiques," 3–8.

53. Buleon, "Political Geography in France," 36.

54. Ibid., 34; Yves Lacoste, *Questions sur géopolitics* (Paris: Librairie Generale Française, 1988), 8. Lacoste's remarks occur within the context of an explanation as to why he chose d'Altdorfer's *La Bataille d'Alexandre* (1529) as the cover of his book on geopolitical problems and issues.

55. See Girot and Kofman, eds., *International Geopolitical Analysis*.

56. Lacoste, "Geography and Foreign Policy," 227.

57. Stanley Hoffman, "An American Social Science: International Relations," *Daedalus* 51 (1977): 41–59.

58. Walker, *Inside/Outside*, 105.

59. See Der Derian and Shapiro, eds., *International/Intertextual Relations*, and Jim George, *Discourses of Global Politics: A Critical (Re)Introduction to International Relations* (Boulder, Colo.: Lynee Rienner, 1994).

60. Richard Ashley, "The Poverty of Neorealism," *International Organization* 38 (Spring 1984): 225–86.

61. Julia Kristeva, "A New Type of Intellectual: The Dissident," in *The Kristeva Reader*, ed. Toril Moi (New York: Columbia University Press, 1986), 292–300; Richard K. Ashley, "Living on Border Lines: Man, Poststructuralism and War," in *International/Intertextual Relations*, ed. Der Derian and Shapiro, 312–13.

62. Kristeva, "New Type of Intellectual," 298; Ashley, "Living on Border Lines," 313.

63. Richard K. Ashley and R. B. J. Walker, eds., "Special Issue: Speaking the Language of Exile: Dissidence in International Studies," *International Studies Quarterly* 34 (September 1990).

64. Richard K. Ashley and R. B. J. Walker, "Introduction: Speaking the Language of Exile: Dissident Thought in International Studies," *International Studies Quarterly* 34 (September 1990): 261.

65. Ashley, "Living on Border Lines," 265–66.

66. Richard K. Ashley and R. B. J. Walker, "Reading Dissidence/Writing the Discipline: Crisis and the Question of Sovereignty in International Studies," in *International Studies Quarterly* 34 (September 1990): 368–70.

67. Ashley and Walker, "Reading Dissidence/Writing the Discipline," 382.

68. In his reading of theoretical discourse on the anarchy *problématique* in IR theory, Ashley suggests that its representations are important "because they replicate on the plane of theory some of the most effective interpretative dispositions and practical orientations by which women and men, statesmen and entrepreneurs go about their business, interpret ambiguous circumstances, impose meaning, discipline and exclude resistant interpretations, and participate in the construction of the conditions, limits, dilemmas, and prevailing ways of knowing and doing that we take to be the familiar truths of global life" (Richard Ashley, "Untying The Sovereign State: A Double Reading of the Anarchy Problematique," *Millennium* 17 [1988]: 227–62, at 228). This is fine as a general justification. However, as he later notes, the representational practices and concerns of those within the national security community, those that seek to actively participate in the making of statecraft, are quite distinct from analysts of the anarchy *problematique* (237).

69. Ashley, "The Poverty of Neorealism," 272.

70. Jim George, in his superb review of the field of IR, astutely identifies this problem when he writes that a genuinely critical perspective rejects "the paradox associated with some postmodernism that, projecting the spirit of tolerant critical theory and practice, engages in clichéd, polemical closure that characterizes so much of the modernism it eschews" (163). As he notes in his conclusion, "the answer to self-affirming grand theory is not self-affirming grand theory" but critical social theory perspectives that reflect the "difficulties, complexities and uncertainties of modern social life rather than the unreflective certitude of some within its elite sectors." See George, *Discourses of Global Politics*, 230.

71. Ashley and Walker, "Reading Dissidence/Writing the Discipline," 395–402.

72. Among the best are David Campbell, *Writing Security* (Minneapolis: University of Minnesota Press, 1992), and Campbell, *Politics without Principle* (Boulder, Colo.: Lynne Rienner, 1993); James Der Derian, *Antidiplomacy;* Michael Shapiro, *Reading the Postmodern Polity;* Bradley Klein, *Strategic Studies and World Order: The Global Politics of Deterrence* (Cambridge: Cambridge University Press, 1994); and Cynthia Weber, *Simulating Sovereignty: Intervention, the State and Symbolic Exchange* (Cambridge: Cambridge University Press, 1995).

73. Richard Ashley, "The Geopolitics of Geopolitical Space: Towards a Critical Social Theory of International Politics," *Alternatives* 12 (1987): 403–34.

74. Richard Ashley, "Untying the Sovereign State," 253.

75. Ashley, "Geopolitics of Geopolitical Space," 423.

76. Walker, *Inside/Outside,* 6.

77. Ibid., 107, 126, where geopolitics is treated as a degenerate form of realism and crude determinism that is an apology for cynicism and political force. This reading of geopolitics is superficial.

78. Dalby, *Creating the Second Cold War.*

79. Dalby, "Critical Geopolitics: Discourse, Difference, Dissent."

80. Richard Ashley, "Foreign Policy as Political Performance," *International Studies Notes* 13 (1987): 51; Campbell, *Writing Security,* 69.

81. Dalby, *Creating the Second Cold War,* 159–60.

6. BETWEEN A HOLOCAUST AND A QUAGMIRE

1. Anthony Lake, "Press Briefing by National Security Advisor Tony Lake and Director for Strategic Plans and Policy General Wesley Clark," May 5, 1994, document 1948, available on the Internet from the information server Almanac@esusda.gov, the Clinton administration's White House documents online service.

2. "Remarks by the President in CNN Telecast of 'A Global Forum with President Clinton,'" May 3, 1988, document 1928, almanac@esusda.gov.

3. The statist conception of responsibility and proximity is not the same as that articulated by Emmanuel Levinas. Engaging with his understandings of ethics and morality—and developments of it—has particular relevance for thinking through a moral response to genocide in the Balkans. See Zygmunt Bauman, *Modernity and Ambivalence* (Cambridge: Polity Press, 1991) and *Postmodern Ethics* (Oxford: Blackwell, 1993); David Campbell, "The Deterritorialization of Responsibility: Levinas, Derrida, and Ethics after the End of Philosophy," *Alternatives* 19 (1994): 455–84, and Critchley, *Ethics of Deconstruction.*

4. President Clinton, "We Force the Spring," *New York Times,* January 21, 1993, A15.

5. Aesthetic televisual visibility refers to how particular events and conflicts lend themselves to the production of dramatic, engaging, and seductive television images, images that have already been anticipated and coded by our thoroughly televisual culture. The images of the dismantling of the Berlin Wall or the nose-cone videos on U.S. smart bombs during the Gulf War are examples. By contrast, certain conflicts and processes, such as low-intensity warfare or the operation of endemic corruption, do not usually make good television.

6. Mackenzie Wark, *Virtual Geography: Living with Global Media Events* (Bloomington: Indiana University Press, 1994), vii.

7. The analysis of contemporary foreign policy issues is always an imperfect exercise. One is dependent upon the public statements of administration policy and upon the ability of the press to provide accounts of policy debates and the policymaking process. While this process is more open in the United States than in other countries, numerous agendas are at work in determining that which becomes known and that which remains secret. "Insider" books like Elizabeth Drew's *On the Edge: The Clinton Presidency* (New York: Simon and Schuster, 1994) are inevitably shaped by their sources and the acceptable tropes of American journalistic culture.

8. This discourse of symbolization has an important psychoanalytic register. Places troped as states of fluidity (morass, whirlpool, and so on) are sites that resist adequate symbolization. They are places of excesses and places of danger, places where the solid forms of *Logos*, Reason, and the Symbolic are threatened with disappearance and engulfment. One should either steer clear of such sites or contain them with solid formations. Luce Irigaray points to a long-standing complicity between rationality and a mechanics of solids, while Cynthia Weber traces the operation of an imperative to dam flows and solidify a geographical region in U.S. foreign policy discourse on the Caribbean. See Luce Irigaray, *This Sex Which Is Not One*, trans. Catherine Porter (Ithaca, N.Y.: Cornell University Press, 1985), 106–18; and Cynthia Weber, "Shoring up a Sea of Signs: How the Caribbean Basin Initiative Framed the U.S. Invasion of Grenada," *Environment and Planning D: Society and Space* 12 (1994): 547–58.

9. The exact quote, from Johnson's discussion with a biographer, is as follows: "If I left the woman I really loved—the Great Society—in order to get involved with that bitch of a war on the other side of the world, then I would lose everything at home. All my programs." See Marilyn Young, *The Vietnam Wars, 1945–1990* (New York: HarperPerennial, 1991), 106.

10. Klaus Theweleit notes the hysterical need on the part of fighting men to win a war to get the lost one back. Loss was femininized by these groups. "Fighting femininity, simultaneously winning the COLD WAR, became

the ways of getting back parts of the Vietnam war as a war won. Women and 'the East' had to give back what had got lost in the swampy triangle of that Asian communist prostitute." See Klaus Theweleit, "The Bomb's Womb and the Genders of War," in *Gendering War Talk,* ed. Miriam Cooke and Angela Woolacott (Princeton: Princeton University Press, 1993), 285; and Susan Jeffords, *The Remasculinization of America* (Bloomington: Indiana University Press, 1989), 144–67.

11. Dan Oberdorfer, "A Bloody Failure in the Balkans: Prompt Allied Action Might Have Averted Factional Warfare," *Washington Post,* February 8, 1993, A1.

12. Richard Schifter, "Human Rights in Yugoslavia," *U.S. Department of State Dispatch* 2, no. 9 (March 4, 1991): 153.

13. Margaret Tutweiler, "U.S. Policy Towards Yugoslavia," *U.S. Department of State Dispatch* 2, no. 22 (June 3, 1991): 395.

14. Secretary Baker, "U.S. Concerns about the Future of Yugoslavia," *U.S. Department of State Dispatch* 2, no. 26 (July 1, 1991): 468.

15. Significance has also been attached to the statement by then NATO commander John Galvin to the Belgrade daily *Politika* that Yugoslavia did not lie within NATO's defense perimeter and that therefore NATO would not intervene in the event of a Yugoslav civil war. See Sabina Petra Ramet, "The Yugoslav Crisis and the West: Avoiding 'Vietnam' and Blundering into 'Abyssinia,'" *East European Politics and Societies* 8, no. 1 (Winter 1994): 197. In his memoirs, Baker describes his day of meetings with the leading Yugoslavian players as one of the most depressing of his time as secretary of state. The leaders, particularly Milosevic and Tudjman, whom he indirectly describes as "bull-headed mountain men" (634), "seemed to be sleep walking into a car wreck" (483). Baker notes that there was never any thought at that time of using U.S. ground troops, for the American people would never have supported it. America's vital interests were not at stake. He explains away the Bush administration's lack of leadership by noting that there was an undercurrent of feeling in Washington "that it was time to make the Europeans step up to the plate and show that they could act as a unified power" (637). Interestingly, in scripting himself as a moral actor who got "worked up" and emotional about Bosnia (although for calculated purpose; no softhead, he), he blames the Europeans for indecision, noting that "in European eyes Sarajevo was becoming Saigon" (637). But those eyes were not European (alone). See James A. Baker III with Thomas M. DeFrank, *The Politics of Diplomacy: Revolution, War and Peace, 1989–1992* (New York: Putnam, 1995), 478–83, 634–51.

16. Secretary Baker, "Violent Crisis in Yugoslavia," *U.S. Department of State Dispatch* 2, no. 39 (September 30, 1991): 723.

17. Richard Johnson, "U.S. Efforts to Promote a Peaceful Settlement in Yugoslavia," *U.S. Department of State Dispatch* 2, no. 42 (October 21, 1991): 782.

18. See David Gompert's largely apologist reading of the Bush administration's response to the Yugoslav crisis, "How to Defeat Serbia," *Foreign Affairs* 73, no. 4 (1994): 30–47. Gompert was the senior director for Europe and Eurasia on the Bush administration's National Security Council Staff. He does note, however, that the U.S. handling of the Yugoslav crisis contradicted and undermined its declared policy regarding the centrality and purpose of NATO in post–Cold War Europe (36).

19. ABC News, *While America Watched: The Bosnia Tragedy,* broadcast March 17, 1994; R. LaMont Jones Jr., "Ex-Envoy Underscores Balkans Fragility," *Pittsburgh Post-Gazette,* March 22, 1994, A4. Zimmermann also records his own experiences during this time in his article "The Last Ambassador," *Foreign Affairs* 74 (1995): 2–21.

20. Roger Thurow and Carla Anne Robbins, "A Costly Restraint: The West's Record in Bosnia Now Leaves Few Choices—All Bad," *Wall Street Journal,* April 30, 1993, A1.

21. Paul Quinn-Judge, "Powell Preparing to Go Public: Politics Would Bring Scrutiny," *Boston Globe,* August 30, 1993, A21.

22. Zimmermann, "Last Ambassador," 15.

23. Oberdorfer, "Bloody Failure."

24. Ibid.

25. Ibid.

26. David Binder, "U.S. Policymakers on Bosnia Admit Errors in Opposing Partition in 1992," *New York Times,* August 29, 1993, 10.

27. It has been claimed that Izetbegovic changed his mind because he was encouraged by U.S. support coming from Zimmermann for a single Bosnia-Herzegovina. Zimmermann, however, denies he encouraged Izetbegovic to back out of the deal, stating that his standing instructions were to support any European Community–brokered agreement acceptable to the three parties. See Binder, "U.S. Policymakers," and Warren Zimmermann, "Bosnia About-Face," *New York Times,* letter, September 30, 1993, A24.

28. Binder, "U.S. Policymakers."

29. Ibid.

30. Editorial, "The Dead of Winter, in Bosnia," *New York Times,* September 28, 1992, A14.

31. "The Candidates on the Balkans," *Los Angeles Times,* August 8, 1992, 8.

32. For the story of Omarska, see Ed Vulliamy, *Seasons in Hell* (London: Simon and Schuster, 1993), and Roy Guttman, *A Witness to Genocide* (New York: Macmillan, 1993).

33. George Bush, "Containing the Crisis in Bosnia and the Former Yugoslavia," *U.S. Department of State Dispatch* 3, no. 32 (August 10, 1992): 617.

34. An alternative perspective came from Boutros Boutros-Ghali who remarked during one of his visits to Sarajevo that what was going on there was a "rich man's war." Among large sections of the Atlanticist intelligentsia, however, the dominant position was neither that of the Bush administration nor that of Boutros Boutros-Ghali but horror that what was going on was occurring in Europe. Sarajevo, it was repeatedly noted, was a European city. See David Rieff, *Slaughterhouse: Bosnia and the Failure of the West* (New York: Simon and Schuster, 1995), 24, 33–52.

35. Acting Secretary Lawrence Eagleburger, "Intervention at the London Conference on the Former Yugoslavia," *U.S. Department of State Dispatch* 3, no. 31 (August 31, 1992): 673. Eagleburger took over from James Baker when the latter left to manage the Bush reelection campaign in August 1992.

36. Even the *Economist* magazine, which took an anti-interventionist stance on the war, argued that the "Balkan feud" school of interpreting Bosnia "underplays the extent to which the soldiers and politicians fighting this war have rational aims, which they are willing to bargain over." "Reinventing Bosnia," *Economist,* August 22, 1992, 14.

37. Rieff, *Slaughterhouse,* 44–45.

38. George Bush, quoted in James Gerstenzang, "Bush Still Balking at Balkans Force," *Los Angeles Times,* August 8, 1992, 18.

39. Michael Gordon, "Powell Delivers a Resounding No on Using Limited Force in Bosnia," *New York Times,* September 28, 1992, A1, A5.

40. Michael Gordon, "Bush Backs a Ban on Combat Flights in Bosnia Airspace," *New York Times,* October 2, 1992, A1, A19.

41. Colin Powell, "Why Generals Get Nervous," *New York Times,* October 8, 1992, A35.

42. Secretary Eagleburger, "Identifying Yugoslav War Criminals," *U.S. Department of State Dispatch* 3, no. 52 (December 28, 1992): 924.

43. Secretary Eagleburger, "The Need to Respond to War Crimes in the Former Yugoslavia," *U.S. Department of State Dispatch* 3, no. 52 (December 28, 1992): 925.

44. "Statement on the Former Yugoslavia." Released by the North Atlantic Council's Ministerial Meeting, Brussels, Belgium, December 17, 1992, *U.S. Department of State Dispatch* 3, no. 52 (December 28, 1992): 929.

45. Oberhofer, "Bloody Failure."

46. Bill Clinton, August 4, 1992, in "The Candidates on the Balkans" (see n. 31).

47. Editorial, "It's Not a Fair Fight," *Los Angeles Times,* October 16, 1992, 6.

48. Thurow and Robbins, "Costly Restraint."

49. Martin Walker, "Bill Clinton's Troublesome Inheritance," *Guardian*, January 30, 1993, 23.

50. The desert is a metaphor for America in Baudrillard. It is a cinematic screen, a vast unencumbered space, and a depthless wasteland that denotes the emptiness that is behind America's radical modernity. While Baudrillard's observations on America are certainly partial and often absurd, they are occasionally provocative. For example, Baudrillard's suggestion that the desert is a primal scene for culture, politics, and sexuality in America does resonate with the way America commanded and fought the Gulf war. The desert during the Gulf War was intensely obscene (in Baudrillard's terms): a place of objectivity, hypervisibility, extermination, and the disappearance of referents (most intensely on the screens of the military watching and war machines). See Jean Baudrillard, *America* (New York: Verso, 1992).

51. Drew, *On the Edge*, 140.

52. Secretary Christopher, "New Steps towards Conflict Resolution in the Former Yugoslavia," *U.S. Department of State Dispatch* 4, no. 7 (February 15, 1993): 81–82. For an account of the formulation of this policy, see Drew, *On the Edge*, 138–48.

53. President Clinton, "Remarks at a Town Meeting in Detroit," *Weekly Compilation of Presidential Documents* 29, no. 6 (February 10, 1993): 178.

54. "The President's News Conference with Prime Minister John Major of the United Kingdom," *Weekly Compilation of Presidential Documents* 29, no. 8 (February 24, 1993): 311.

55. President Clinton, "Remarks on Mayoral Support for the Economic Plan and on Exchange with Reporters," *Weekly Compilation of Presidential Documents* 29, no. 9 (March 5, 1993): 361.

56. Drew quotes one "high-level official" at these meetings who described them not as policymaking but as "group therapy—an existential debate over what is the role of America, etc." Drew, *On the Edge*, 150.

57. Tim Weiner, "State Dept. Official Says U.S. Ignores 'Genocide' in Bosnia," *New York Times*, February 4, 1994, A4.

58. President Clinton, "U.S. Holocaust Museum Dedicated," *U.S. Department of State Dispatch* 4, no. 19 (May 10, 1993): 323.

59. Lexington, "A Puzzled People," *Economist*, May 8, 1993, 36. Drew writes that Clinton aides reported that he was "deeply affected, and influenced in his position on Bosnia, by the ceremonies surrounding the opening of the Holocaust museum"; *On the Edge*, 153.

60. *Crossfire*, "Never Again?" April 22, 1993, Cable News Network.

61. Ibid.

62. See Matthew Vita, "Hawks, Doves in D.C. Trade Places amid Flutter of Debate over Bosnia," *Atlanta Journal and Constitution,* April 29, 1993, A14; *Crossfire,* "Turning Up the Heat in Bosnia," April 27, 1993, Cable Network News.

63. Michael Ross, "Clinton Urged to Take Case for Bosnia Moves to Public," *Los Angeles Times,* May 5, 1993, 7. For a reading of the debate among the public, which also oscillated between Holocaust and Vietnam analogies, see Sara Fritz, "Hawks, Doves among Public Switch Sides," *Los Angeles Times,* May 9, 1993, 1. See also Martin Walker, "The Debate in the U.S.: Clinton Hesitates as a Nation Divided Shies Away from War," *Guardian,* April 29, 1993, 13.

64. UNSC Resolution 824 (May 6, 1993), *U.S. Department of State Dispatch* 4, no. 20 (May 17, 1993): 348.

65. Secretary Christopher, "U.S. Consultations with Allies on Bosnia-Herzegovina," *U.S. Department of State Dispatch* 4, no. 19 (May 10, 1993): 321.

66. Drew, *On the Edge,* 155.

67. Secretary Christopher, "Testimony before Congress," *U.S. Department of State Dispatch* 4, no. 22 (May 31, 1993): 395.

68. Doyle McManus, "Agony at the Top: Bosnia May Be Clinton's Vietnam," *Los Angeles Times,* April 29, 1993, 1. This article notes that "Vietnam and the Holocaust are the twin phantoms that haunt the Clinton Administration's debate over what to do in the Bosnian highlands."

69. American Survey, "Into Bosnia?" *Economist,* May 15, 1993, 25.

70. Robert Kaplan, *Balkan Ghosts: A Journey through History* (New York: St. Martin's Press, 1993). Kaplan's book is a deeply Eurocentric one that writes the Balkans as Europe's original Third World, an Africa on the European continent, its premodern, savage Other. According to Kaplan, the Balkans are the crucible of the twentieth century, the place where terrorism and genocide first became tools of policy, the homeland of Nazism. Kaplan's whole project is framed as an adventure journey through an exotic past-as-present. It sites the region from a gaze anchored in medieval churches and monasteries and guided by the cites of three Western authorities (Rebecca West, John Reed, and C. L. Sulzberger). The result is yet another postmodern "colonial *New York Times*" correspondent's perspective on a region of the world, this one unconsciously reveling in the raw hatred and slaughter stories of a region supposedly overwhelmed by its history. Missing is any sustained discussion of power and class. Culture and the weight of history determine all. This whole "hotel balcony gaze," Western-correspondent genre of writing place and global space — the genre of Thomas Friedman, Steven Kinzer, and others — deserves a lot more critical deconstruction than it gets from geographers.

71. Drew, *On the Edge,* 159.

72. Secretary Christopher, "Announcement of the Joint Action Program on the Conflict in Bosnia," *U.S. Department of State Dispatch* 4, no. 21 (May 24, 1993): 368–69.

73. Art Pine, "U.S., Allies Forge Strategy on Bosnia to Contain Fighting," *Los Angeles Times,* May 23, 1993, 1. Clinton's remarks were made in New Hampshire on May 22, 1993.

74. Editorial, "The Bosnia Decision," *Economist,* May 29, 1993, 14–15.

75. Drew, *On the Edge,* 160.

76. Clinton was moved by television images of the siege of Sarajevo (in July and September) sufficiently to call for new policy options (worked on by Lake). However, Clinton never followed through. Drew writes: "He was torn between his emotional responses and his domestic imperative, which always prevailed. His emotional instincts were more than once checkmated by his political instincts.... In the absence of steady leadership, his policy was all over the place, and his policymakers were in a state of confusion over what he wanted—and vied for his mind." Drew, *On the Edge,* 283.

77. Ibid., 275–84.

78. Anthony Lake, "From Containment to Enlargement," *U.S. Department of State Dispatch* 4, no. 39 (September 27, 1993): 663.

79. Ibid., 662.

80. President Clinton, "Confronting the Challenges of a Broader World," *U.S. Department of State Dispatch* 4, no. 39 (September 27, 1993): 652. Anthony Lake outlined the "tough questions" the United States would ask before committing itself to UN peacekeeping operations in "The Limits of Peacekeeping," Opinion Editorial, *New York Times,* February 6, 1994, 17.

81. Secretary Christopher, "The Strategic Priorities of American Foreign Policy," *U.S. Department of State Dispatch* 4, no. 47 (November 22, 1993): 797.

82. Elaine Sciolino and Douglas Jehl, "As U.S. Sought a Policy, the French Offered a Good Idea," *New York Times,* February 14, 1994, A1, A6.

83. "Statement by the President," February 9, 1994, document 1410, almanac@esusda.gov.

84. In praising the statesmanship of Croat president Franjo Tudjman, Clinton was praising a man who was directly responsible for fanning the flames of the murderous nationalism that swept across Yugoslavia in 1992. Tudjman not only presided over the creation of an exclusivist, anti-Serb Croat state that adopted many of the symbols of the Nazi-sponsored Croat state during World War II, but proposed to Milosevic that they both divide Bosnia in 1992. See Misha Glenny, *The Fall of Yugoslavia* (London: Penguin, 1992).

85. The Serb offensive against Gorazde, the first significant attack on a UN-NATO-designated "safe area" since Sarajevo, began on March 30. Five days passed before President Clinton spoke out against the attack. Twelve more days passed before he sat down with his national security advisers on Bosnia. It was not until April 22 that the ultimatum against the Serbs was issued. To counteract criticism of the president's disengagement from foreign affairs, the White House circulated a four-page document with the title "Draft Chronology of the President's Foreign Policy Engagement, April 4–21." See Douglas Jehl and Elaine Sciolino, "Sharper Focus: Genesis of Clinton's Hard Line," *New York Times,* April 24, 1994, A14.

86. In fact, politicians and the media tend to use the terms "safe area" and "safe haven" interchangeably, as if they were the same.

87. On General Rose's role, see Rieff, *Slaughterhouse,* 187–89, 206–7, and ABC News, *The Peacekeepers: How the United Nations Failed in Bosnia,* broadcast April 24, 1995.

88. Clinton first revealed that the United States and France were working on a map in an address to the French Assembly. See "Remarks by the President in Address to the National Assembly," June 7, 1994, document 2148, almanac@esusda.gov; Elaine Sciolino, "U.S. Backs Bid to Press Peace on Bosnia Foes," *New York Times,* June 9, 1994, 1.

89. Jenonne Walker, "Avoiding Risk and Responsibility: The United States and Eastern Europe," *Current History* 91 (1992): 364–68.

90. "United Nations Peacekeeping," *Economist,* June 25, 1994, 19; and State Department Briefing by Christine Shelly, June 10, 1994, available through the Federal News Service. The word "genocide" was not so much banned as prohibited from use in isolation as a pure objective descriptor; spokespersons were to use the phrase "acts of genocide" to convey the conditionality and impreciseness of situations such as Rwanda. This led to the absurd situation of the State Department recognizing that "acts of genocide" had taken place in Rwanda, but Rwanda as a whole was not necessarily the site of a genocide. Restricting the term in this manner was an attempt to contain not only an obvious moral imperative but the legal obligation of the United States and other countries to establish an international tribunal to prosecute the guilty as a signatory of the 1948 Genocide Convention. One of the most outspoken critics of the Clinton administration's stance on Bosnia in the House of Representatives, Frank McCloskey (D-Indiana), constantly complained that the administration was ignoring genocide in Bosnia. In testimony before the House in November 1993, Warren Christopher responded to McCloskey's criticism by declaring, "Your very strong feelings on this subject have affected your judgment." Speaking in terms of genocide was to speak the language of emotion, not the lan-

guage of foreign policy strategy. The exchange demonstrates how substantive moral reasoning was always to be subordinated to narrow, bureaucratic, instrumental reasoning in the case of Bosnia.

91. Yet the Serbian drive toward homogeneous statehood was also a reaction to a certain modernity, a postcommunist postmodernity of uncertain economic prospects, speed, and tumultuous change. In such conditions, as David Harvey has argued, mythological and aesthetic visions of the state, territory, and identity tend to emerge as attempts to create stability amid periods of "creative destruction" and intensified time-space compression. See David Harvey, *The Condition of Postmodernity* (Oxford: Blackwell, 1989).

92. Bauman, *Modernity and Ambivalence,* 26–39.

93. Michael Ignatieff points out how the "misery of the Balkan people does not derive from their home-grown irrationality, but from the pathetic longing to be good Europeans, that is, to import the West's most murderous ideological fashions." See Ignatieff, "The Balkans Tragedy," *New York Review of Books* 40, no. 9 (May 13, 1993): 3–5.

94. See Gearóid Ó Tuathail, "An Anti-Geopolitical Eye: Maggie O'Kane in Bosnia, 1992–93," *Gender, Place and Culture* 3, no. 2 (1996): 171–85.

95. It can be argued, following Baudrillard, that reading "Bosnia" as a "Holocaust" is a fictionalization of the conflict, the production of "Bosnia" as hyperreal, as a "Holocaust" simulation. Writing "Bosnia" as a "Holocaust" thus denotes the loss of our ability to understand the violence habitually generated by our modernity. It marks the triumph of fictitious evils from history, fables that send us off hunting for Nazis and losing the reality of postcommunism in the process. Like many of Baudrillard's points, however, this argument is overstated. It imputes a logic to the operation of historical fables they do not have. It also encourages an arrogant dismissal of the work of the many intellectuals who struggle to render the actuality of the horror of Nazism/Auschwitz/Bosnia intelligible. See Stjepan Mestrovic, *The Balkanization of the West* (London: Routledge, 1994), 27–54.

96. This does not mean that I believe the United States is pro-Serbian. I disagree with Stjepan Mestrovic, who argues, in *The Balkanization of the West,* that the West is unconsciously glad that Serbia is oppressing the Muslims and that it obtains vicarious gratification from watching Serbian atrocities on television. He reads the Balkans conflict by crudely deploying the concepts of "modernism" and "postmodernism" to argue that postcommunism is characterized not by the triumph of capitalist modernity but by the ascendancy of a postmodern narcissism (fundamentalism, nationalism, and Balkanization). The West's response to the conflict is characterized as "the simulation of concern, the fiction of humanitarianism, and make-believe diplomacy" (88). Mestrovic's argument stereotypes modernity, post-

modernity, and the West; it attributes a logic and coherence to each of these that is unjustified and essentializing. Parts of his book are little more than apologism for Croatia and Franjo Tudjman.

97. See Rieff, *Slaughterhouse*, 162–89.

98. "From the perspective of the 'rational order,' morality is and is bound to remain *irrational*" (Bauman, *Postmodern Ethics*, 13). Bauman develops this postmodern vision of morality from his reading of Levinas supplemented by Derrida. The denunciation of humanitarianism echoes Levinas's denunciation of humanism because it is not sufficiently human. See Campbell, "The Deterritorialization of Responsibility," 462.

99. Hans Mommsen, quoted in Bauman, *Modernity and Ambivalence*, 192.

7. VISIONS AND VERTIGO

1. Richard Barnet and John Cavanagh, *Global Dreams: Imperial Corporations and the New World Order* (New York: Simon and Schuster, 1994), 22.

2. James Kurth, "The *Real* Clash," *National Interest* 37 (Fall 1994): 3–4.

3. For a discussion of this post–Cold War disorientation, see Gearóid Ó Tuathail and Timothy W. Luke, "Present at the (Dis)Integration: Deterritorialization and Reterritorialization in the New Wor(l)d Order," *Annals of the Association of American Geographers* 84 (1994): 381–98.

4. Zbigniew Brzezinski, *Out of Control: Global Turmoil on the Eve of the Twenty-First Century* (New York: Macmillan, 1993), ix–x.

5. "Geopolitical Vertigo and the U.S. Role," *New Perspectives Quarterly*, Summer 1992, 5–33.

6. Harvey, *The Condition of Postmodernity*.

7. Jean-François Lyotard, *The Postmodern Condition* (Minneapolis: University of Minnesota Press, 1984); Gianni Vatimo, *The Transparent Society* (Baltimore, Md.: Johns Hopkins University Press, 1992), 5.

8. Wark, *Virtual Geography*, x–xiv.

9. Jean Baudrillard, *The Ecstasy of Communication* (New York: Semiotext[e], 1988), 12.

10. "The *mise en abŷme* is...a kind of complication or frenzy in the systems of representation. There is a doubling, a multiplication of the system. But in my opinion it is a kind of impasse or extremity of a system of representation, which becomes more sophisticated, certainly, but which loses the thread of representation." Jean Baudrillard, *Baudrillard Live: Selected Interviews*, ed. Mike Gane (London: Routledge, 1993), 83–84.

11. Douglas Kellner, ed., *Baudrillard: A Critical Reader* (Cambridge, Mass.: Blackwell, 1994), 13. It should be added, however, that theorists like

Donna Haraway and Timothy Luke look to science fiction for a means of describing the new hybrids of postmodern life. See Haraway, *Simians, Cyborgs, and Women* (New York: Routledge, 1991), and Timothy W. Luke, "At the End of Nature: Cyborgs, Humachines and Environments in Postmodernity," *Environment and Planning A* 28 (1996).

12. Useful critiques of Baudrillard include Douglas Kellner, *Jean Baudrillard: From Marxism to Postmodernism and Beyond* (Stanford, Calif.: Stanford University Press, 1989); certain of the essays in Chris Rojek and Bryan Turner, eds., *Forget Baudrillard* (London: Routledge, 1993); and Kellner, ed., *Baudrillard: A Critical Reader*.

13. The point is not that Baudrillard is simply against intellectuals (a point made by Rojek) but that his method actively inhibits attempts to understand the exercise of power in the postmodernity. Zygmunt Bauman sums up the debate over the possibility of critique by sensibly noting that the "likelihood of critique is but loosely, if at all, related to the philosophical elegance of the proof of its legitimacy." See Zygmunt Bauman, "The Sweet Scent of Decomposition," and Chris Rojek, "Baudrillard and Politics," in *Forget Baudrillard*, ed. Rojek and Turner, 44 and 111.

14. Jean Baudrillard, *Simulations* (New York: Semiotext[e], 1983), 2.

15. The notions of deterritorialization and reterritorialization are associated with the work of Gilles Deleuze and Félix Guattari. They describe the ways in which capitalism deterritorializes desire by subverting the traditional codes that limit and control social relations of production. The "creative destruction" of capitalism scrambles the previously fixed order of class, kinship, belief, customs, and space, thus producing social flux. In so doing, it also reterritorializes desire by organizing bits and pieces of the old into a new schizophrenic order. My interest is less in the psychoanalytic dimensions of this double dynamic than in its materialist and geographical dimensions. See Gilles Deleuze and Félix Guattari, *Anti-Oedipus: Capitalism and Schizophrenia*, trans. Robert Hurley, Mark Seem, and Helen Lane (Minneapolis: University of Minnesota Press, 1982), and *A Thousand Plateaus: Capitalism and Schizophrenia*, trans. Brian Massumi (Minneapolis: University of Minnesota Press, 1987).

16. See Richard O'Brien, *Global Financial Integration: The End of Geography* (New York: Council on Foreign Relations Press, 1992); Stuart Corbridge, Ron Martin, and Nigel Thrift, eds., *Money, Power and Space* (Oxford: Blackwell, 1994). As Martin and Thrift make clear in their essays, the globalization of financial markets does not imply the "end of geography." If anything, it has accentuated the importance of place (as centers of market interpretation and social interaction) and further polarized space into a capital-rich North and an indebted South.

17. Susan Strange, *Casino Capitalism* (Oxford: Blackwell, 1986). Among other things, casino capitalism has put the meaning of money increasingly in doubt.

18. Robert Reich, *The Wealth of Nations* (New York: Knopf, 1992).

19. As a reaction to the so-called counterculture of the 1960s, it can be argued that neoconservativism has always been about the reterritorialization of the identity of "America" around mythical ideas of "tradition" and "authority" in reaction to global economic transformations and the "breakdown" of Cold War society. See David Campbell, *Writing Security*, 181; Timothy Luke, *Screens of Power*, 61–86; and John Ehrman, *The Rise of Neoconservatism: Intellectuals and Foreign Affairs, 1945–1994* (New Haven, Conn.: Yale University Press, 1995).

20. For more on the society of security, see Campbell, *Writing Security*, 166–87.

21. On the significance of the jeremiad as a mode of narration in U.S. foreign policy, see Campbell, *Writing Security*, 33–34, 180–81. Campbell makes the important point that "the cold war was both a struggle which exceeded the military threat of the Soviet Union, and a struggle into which any number of potential candidates—regardless of their strategic capacity to be a threat—were slotted as a threat." The cold war, in other words, is deeper than the Cold War. Indeed, recalling that the phrase "cold war" was coined by a fourteenth-century Spanish writer to represent the persistent rivalry between Christians and Arabs, Campbell argues that "the sort of struggle the phrase denotes is a struggle over identity: a struggle that is not context-specific and thus not rooted in the existence of a particular kind of Soviet Union" (34).

22. Andrew Cockburn, "The Defense Intellectual: Edward N. Luttwak," *Grand Street* 6 (Spring 1987): 161–74.

23. Edward Luttwak, *The Grand Strategy of the Roman Empire* (Baltimore, Md.: Johns Hopkins University Press, 1976); Luttwak, *The Grand Strategy of the Soviet Union* (New York: St. Martin's Press, 1983); and Luttwak, *The Pentagon and the Art of War* (New York: Simon and Schuster, 1984).

24. Edward N. Luttwak, "From Geopolitics to Geo-Economics," *National Interest* 20 (1990): 17–24, quote at 17.

25. Edward Luttwak, *The Endangered American Dream* (New York: Simon and Schuster, 1993).

26. Most, but not all, of these literatures concerned themselves with the U.S.-Japan relationship. For a review, see Gearóid Ó Tuathail, "Japan as Threat: Geo-Economic Discourses on the USA-Japan Relationship in U.S. Civil Society, 1987–1991," in *The Political Geography of the New World Order*, ed. Colin Williams (London: Belhaven, 1993), 181–209.

27. This image of the United States as a fraying superpower is one that recurs throughout Tom Clancy's novel *Debt of Honor* (New York: Putnam, 1994). A right-wing novelist whose works of fiction are as much a part of the Cold War society of security as any factual discourse, Clancy's novels (which usually become Hollywood movies) are suggestive barometers of the structure of feeling within the national security state. In *Debt of Honor,* which features a sneak Japanese geo-economic and naval attack on the United States, that structure of feeling is one of resentment at the relative decline of U.S. power and influence in the world. One passage recalls Luttwak's seemingly trivial comments about dilapidation: "She walked to the elevator and rode to the fifth floor of the Old Headquarters Building, now almost forty years of age and showing it. The corridors were dingy, and the paint on the drywall panels faded to a neutral, offensive yellow. Here, too, the mighty had fallen, and that was especially true of the Office of Strategic Weapons Research. Once one of CIA's most important agencies, OSWR was now scratching for a living" (216).

28. For an account of Kissinger's role in the CFIA, see Isaacson, *Kissinger: A Biography* (New York: Simon and Schuster, 1992), 94–107.

29. "The real challenge which communists pose to modernizing countries is not that they are so good at overthrowing governments (which is easy), but that they are so good at making governments (which is far more difficult). They may not provide liberty, but they do provide authority; they do create governments that can govern" (Samuel Huntington, *Political Order in Changing Societies* [New Haven, Conn.: Yale University Press, 1968], 8). On the place of Huntington's ideas within the ideology of liberal modernization in U.S. foreign policy, see Bradley Klein, *Strategic Studies and World Order: The Global Politics of Deterrence* (Cambridge: Cambridge University Press, 1994), 99–101.

30. Samuel Huntington, "The United States," in *The Crisis of Democracy,* ed. Michael Crozier, Samuel Huntington, and Joji Watanuki (New York: New York University Press, 1975), 113.

31. Huntington's views on these issues are expressed in the interview "A Better America: Ideas for the President from Experts; Arms Control Must Not Be 'Oversold,'" *U.S. News and World Report,* January 7, 1985, 75.

32. Samuel Huntington, "America's Changing Strategic Interests," *Survival* 33 (January/February 1991): 3.

33. Shintaro Ishihara, *The Japan That Can Say No,* trans. Frank Baldwin (New York: Simon and Schuster, 1989).

34. Samuel Huntington, "The Economic Renewal of America," *National Interest* 28 (Spring 1992): 14–18.

35. Samuel Huntington, "Why International Primacy Matters," *International Security* 17, no. 4 (Spring 1993): 76.

36. Samuel Huntington, "The Clash of Civilizations?" *Foreign Affairs* 72 (Summer 1993): 22–49; "The Coming Clash of Civilizations or, the West against the Rest," *New York Times,* June 6, 1993, E19; "If Not Civilizations, What? Paradigms of the Post–Cold War World," *Foreign Affairs* 72 (November/December 1993); and *The Clash of Civilizations: The Debate* (New York: Council on Foreign Relations, 1993).

37. Huntington, "If Not Civilizations, What?," 187.

38. Ibid., 186, 191. Huntington appropriates the term from William James, who used it to describe thought without concepts, theories, models, and paradigms.

39. Huntington, "The Clash of Civilizations?" 31. Huntington's map is strange in that it arbitrarily chooses 1500 as a dividing time for a Velvet Curtain that is only emerging in 1990, and it lumps both Orthodox Christianity and Islam together on the same side of the divide.

40. Fouad Ajami, "The Summoning," *Foreign Affairs* 72 (September/October 1993): 7.

41. Richard Rubenstein and Jarle Crocker, "Challenging Huntington," *Foreign Policy* 96 (Fall 1994): 115. They sum up Huntington's argument thus: "The old Cold War is dead, he loudly declares. Then—*sotto voce*—Long live the new Cold War!" (117).

42. Interestingly, this is now conceded by many conservatives. See Owen Harries, "The Collapse of 'The West,'" *Foreign Affairs* 72 (July/August 1993): 41–53, and Kurth, "The *Real* Clash." In his sycophantic essay on Huntington's thesis, Kurth writes, "The tale of the decline of 'Western civilization' as a term is part of the longer tale of the decline of Western civilization itself" (10). He connects this to the transformation from an industrial to a postindustrial economy and from an international economy to a global one, which means that "the most advanced countries are becoming less modern (i.e. postmodern), while the less advanced countries are becoming more modern" (12). The West, he suggests, is now being "deconstructed" by a multicultural coalition and a feminist movement that is marginalizing Western civilization in the universities and media of America. At the very moment of its greatest triumph, Western civilization is becoming non-Western because it has become global and because it has become postmodern. The real clash, for Kurth, is the clash between the West and the post-West, or the modern and the postmodern within the West itself.

43. Huntington, "The Clash of Civilizations?" 39. At another point he writes, "Japan has established a unique position for itself as an associate member of the West: it is in the West in some respects but clearly not of the West in important dimensions" (45). This locational uniqueness of Japan makes it potentially even more of a threat in that it is not of the West yet it

operates within its territorial economies and within its community of confidence. The potential of Japan to divide the West and undermine its security from within is consequently all the greater. Tom Clancy plays upon this important imaginative fear in *Debt of Honor* by having American lobbyists and Japanese corporations attempt to subvert U.S. democracy. See also Pat Choate, *Agents of Influence* (New York: Knopf, 1990).

44. That Huntington supports any of these ideals is, of course, contestable, given his historical support for brutal and repressive Third World military governments.

45. Huntington, "If Not Civilizations, What?" 190.

46. Kurth, "The *Real* Clash," 10.

47. John Pickles, ed., *Ground Truth: The Social Implications of Geographic Information Systems* (New York: Guilford, 1995).

48. Paul Virilio and Sylvère Lotringer, *Pure War,* trans. Mark Polizzotti (New York: Semiotext[e], 1983), 115.

49. See Robert Kaplan, "The Coming Anarchy," *Atlantic Monthly* 273 (February 2, 1994): 44–76.

50. This estimate is from the Russian Ministry of Internal Affairs and is for the year 1993. See Stephen Handelman, "The Russian 'Mafiya,' " *Foreign Affairs* 73 (March/April 1994): 83–96; Seymour Hersh, "The Wild East," *Atlantic Monthly,* June 1994, 61–86. The discursive reading of the Russian mafiya threat is interesting. Images of American gangsterism from the 1930s dominate. Hersh quotes one Bush administration official who declared: "The chaos scares me. Here we have a 1930s situation in Chicago, except that Al Capone has access to nuclear weapons" (79). The fact that Stephen Handelman's article on the Russian mafiya in *Foreign Affairs* is illustrated with a photographic still of James Cagney from the U.S. gangster movie *Public Enemy* is a remarkable testimony to how the conceptualization of this and other foreign policy issues is contaminated by cinematic signs of signification.

51. The notion of wild and tame zones is discussed at greater length in Ó Tuathail and Luke, "Present at the (Dis)Integration." See also Max Singer and Aaron Wildavsky's *The Real World Order: Zone of Peace/Zones of Turmoil* (Chatham, N.J.: Chatham House, 1993).

52. See Paul L. Knox, "Capital, Material Culture, and Socio-Spatial Differentiation," in *The Restless Urban Landscape,* ed. P. L. Knox (Englewood Cliffs, N.J.: Prentice-Hall, 1993), 1–3; and Knox, "The Stealthy Tyranny of Community Spaces," *Environment and Planning A* 26 (1994): 170–73.

53. See Timothy W. Luke, "Worldwatching at the Limits of Growth," *Capitalism, Nature, Socialism* 5, no. 2 (1995): 43–63; and Simon Dalby, "The Politics of Environmental Security," in *Green Security or Militarized Environment?* ed. J. Kakonen (Aldershot: Dartmouth, 1994), 25–53.

54. President Clinton, "Advancing a Vision of Sustainable Development," *U.S. Department of State Dispatch* 5, no. 29 (July 18, 1994): 477–79.

55. As we have noted, President Clinton declared in his inaugural address that there is no longer any distinction between domestic and foreign policy. His transnational liberalism has been distinguished by what John Stremlau has described as a "big emerging markets" (BEM) strategy focused on China (including Taiwan and Hong Kong), India, Indonesia, Brazil, Mexico, Turkey, South Korea, South Africa, Poland, and Argentina. These "regional economic drivers" containing over one-half of the world's population are key to the Clinton doctrine of creating a world community committed to unrestricted global trade flows. See John Stremlau, "Clinton's Dollar Diplomacy," *Foreign Policy* 97 (Winter 1994): 18–35.

56. See Paul Kennedy, *Preparing for the Twenty-First Century* (New York, Random House, 1993); "Overpopulation Tilts the Planet," *New Perspectives Quarterly* 11, no. 4 (Fall 1994): 4–6; Matthew Connelly and Paul Kennedy, "Must It Be the Rest against the West?" *Atlantic Monthly* 274 (December 6, 1994): 61–83; and Virginia Abernethy, "Optimism and Overpopulation," *Atlantic Monthly* 274 (December 6, 1994): 84–111.

57. Mann, *The Sources of Social Power,* vol. 2, *The Rise of Classes and Nation-States.*

58. Ibid., 440.

Index

Dr. Gearóid Ó Tuathail (Gerard Toal) is associate professor of geography at Virginia Tech. He received his Ph.D. from the Department of Geography at Syracuse University, his M.A. from the University of Illinois, and his B.A. (Hons.) from St. Patrick's College, Maynooth, Ireland. He has taught at Syracuse University, the University of Minnesota, and the University of Liverpool, and has published numerous articles on geographical history, U.S.-Japan relations, and contemporary global geopolitical change. He currently writes the annual progress reports on political geography for the journal *Progress in Human Geography*. In 1993 he won an award from the monthly academic journal *Environment and Planning A* for the best paper published in the journal in 1992.

Printed in the United States
by Baker & Taylor Publisher Services